Editor
Gisela Lee

Editorial Manager
Karen J. Goldfluss, M.S. Ed.

Editor-in-Chief
Sharon Coan, M.S. Ed.

Illustrator
Blanca Apodaca

Cover Artist
Barb Lorseyedi

Art Coordinator
Denice Adorno

Creative Director
Elayne Roberts

Imaging
James Edward Grace

Product Manager
Phil Garcia

Publishers
Rachelle Cracchiolo, M.S. Ed.
Mary Dupuy Smith, M.S. Ed.

PROBLEM-SOLVING MATH JOURNALS FOR PRIMARY STUDENTS

Author

Mary Rosenberg

Teacher Created Materials, Inc.
6421 Industry Way
Westminster, CA 92683
www.teachercreated.com

ISBN-1-57690-963-8

Table of Contents

Introduction to Book

Problem-Solving Math Journals for Primary Students is designed to develop logical and mathematical thinking skills in students and to cover the basic math concepts and skills presented in primary-level grades.

The Teacher's Supplement provides Math Practice Assignment Sheet Sets, Math Journal Review Question Sets, and several reproducible pages in the Classroom Tools section for the students to use at home and/or at school.

The Math Practice Assignment Sheet Sets are a way to review/preview math skills at home. Each assignment sheet has 24 different math questions covering various skills and concepts. The questions cover basic math skills, as well as ask some challenging questions for the more capable students. The assignment sheets can be used as a homework supplement. They may be distributed at the beginning of each month, or you may wish to make each student his or her own packet of all the practice sheets.

The Math Journal Review Question Sets are designed to be used after some math concepts have been discussed and practiced. The review pages are just what their name implies—they review the math skills that were previously covered. The review pages provide a quick and easy way to assess the math strengths and weaknesses of each student.

The Teacher's Supplement also has a set of Classroom Tools (reproducible masters) for teachers to use in the classroom to assist students in solving the problems. These pages can be photocopied onto cardstock or construction paper and laminated or put in a plastic sleeve for durability, if desired. Make a set of these for the students to keep at school and another set to keep at home (if needed).

Introduction to Book *(cont.)*

There are many journal practice sheets in the Student's Problem Solving Math Journal section that explore a variety of math concepts and skills—counting; addition and subtraction; time and money; numbers to 100; and place value. There is a journal practice sheet for each day of school.

Each practice sheet begins with a warm up. The warm up is designed to review or preview a new skill and to get the students' minds into a math mode. The middle part of each practice sheet asks the students to solve a word problem. The student might have to add manipulatives or counters provided on pages 251–256 to solve some of the problems. (*Note:* There are usually more manipulatives or counters provided than will necessarily be used for some problems.) Similarly, the student might have to draw a picture to show how he or she solved the word problem. After coming up with a solution, the student writes a math problem (if applicable) and writes a sentence describing the solution. Before making copies to distribute to students, you may want to cover the answer key (if given) at the bottom of the page.

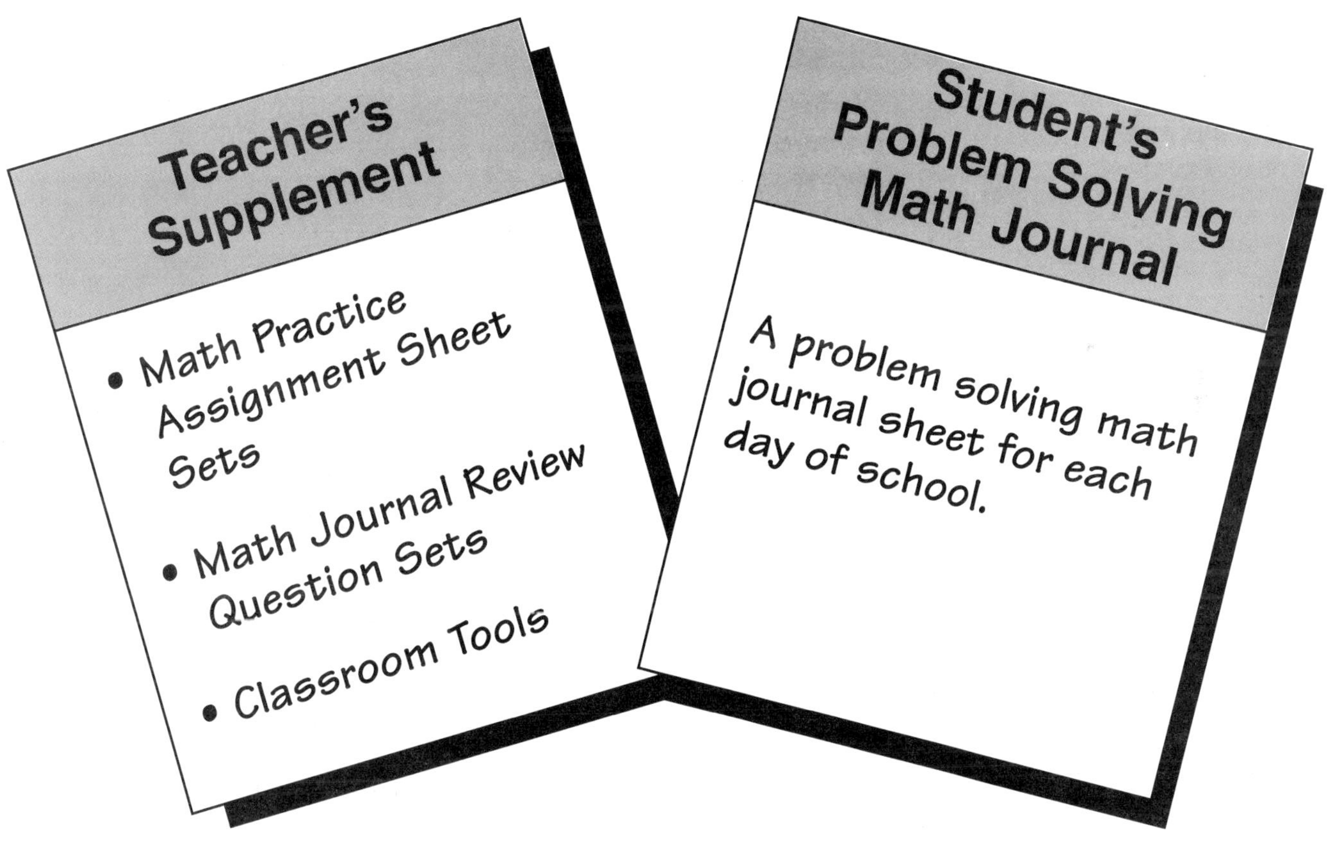

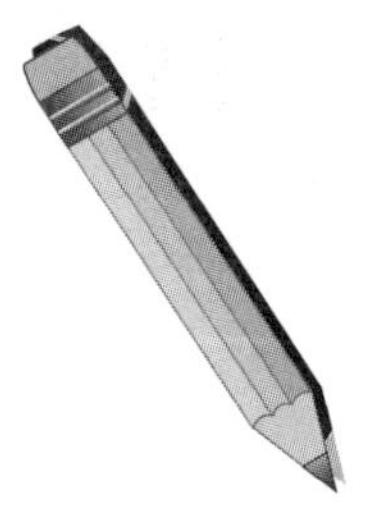

TEACHER'S SUPPLEMENT

- **Math Practice Assignment Sheet Sets**
- **Math Journal Review Question Sets**
- **Classroom Tools**

Introduction to Teacher's Supplement

The Teacher's Supplement provides many resources for teachers to use. Included are sets of math practice assignment sheets which are reproducible pages that students can use at home (as a homework supplement). Each box contains a problem for the student to solve that are similar to the types of problems that he or she will encounter in the journal pages. The student can work out the problem on the page itself or use another sheet of paper to solve the problem. Encourage each student to attempt at least 18 of the 24 problems for each set of problems and show his or her work on how they solved them.

The Math Journal Review Question Sets should be used following the completion of correlating concepts in the journal pages. [**Note:** It is at the teacher's discretion to decide when to use the review questions (i.e., weekly, monthly, etc.)]. The review pages will help teachers assess the strengths and weaknesses for each student.

The Teacher's Supplement also provides Classroom Tools, a set of reproducible teaching aides which include the following:

- numbers and number words
- addition chart
- hundreds chart
- inch rulers
- centimeter rulers
- number cards
- number lines
- place value counters for tens and ones
- place value charts
- math symbols
- number puzzle board
- money
- clock pattern
- shape patterns

These pages can be reproduced onto cardstock or index paper so teachers can make a classroom set. Teachers may also wish to reproduce them individually and distribute to students when they correspond with the math lesson.

Math Practice Assignment Sheet Set #1

Name: ______________________ **Due Date:** ______________

Directions: During this month complete at least 18 of the questions. After completing the questions, have your teacher or parent initial the work sheets. Some of the questions will need to be completed on a separate sheet of paper. Staple those pages to the back of the work sheets. Return the work sheets and all attached pages on the assigned due date.

Draw a picture of your family—don't forget to include yourself! How many people are in your family? **1**	Draw a picture and write a math problem. Muriel has 2 orange crayons and 1 brown crayon. How many crayons does Muriel have in all? **2**	Candace planted 3 pumpkin seeds. Each seed can grow two pumkins. What is the largest number of pumpkins that can grow from the 3 pumpkin seeds? **3**
Draw a picture and write a math problem. Ivan has 4 pencils. Each pencil has an eraser. How many erasers does Ivan have in all? **4**	Draw a picture and write a math problem. On Monday, the bees saw 3 bears. On Tuesday, the bees saw 2 bears. How many bears did the bees see in all? **5**	How many teeth have you lost so far this year? Make a picture showing your smile and the number of teeth you have lost. **6**
The tooth fairy leaves 1¢ for each tooth she collects. How much money has she left for you? **7**	Write your first name. How many letters are in your first name? **8**	Write your last name. How many letters are in your last name? **9**
Make a chart of all the names of your family members. How many letters are in each name? Whose name is the longest? Shortest? **10**	How many legs are there on a snake? Make a picture to show your answer. **11**	Draw a picture showing the animals in the correct order. The bee is first. The ladybug is last. The grasshopper is in between the bee and the ladybug. **12**

Math Practice Assignment Sheet Set #1 *(cont.)*

Name: ______________________________ **Due Date:** ________________

Draw a big circle on a piece of paper. Draw 3 happy faces inside the circle. Draw 4 happy faces outside the circle. How many happy faces did you make in all? **13**	Draw a picture and write a math problem. June saw 8 sheep on Tuesday and 2 sheep on Friday. How many sheep did June see in all? **14**	Draw a picture and write a math problem. Amy picked up a total of 7 golf balls. At the first hole, she picked up 4 golf balls. How many golf balls did Amy pick up at the second hole? **15**
Draw a picture and write a math problem. There are 6 students in a class. There are 3 girls and the rest are boys. How many boys are in this class? **16**	Jessica's classroom number is 3. Her friend's classroom is 4 doors away. In what classroom number is Jessica's friend in? **17**	Write the numbers from 0 to 10. Draw a circle around your favorite number. **18**
Write the numbers from 10 to 0. Draw a square around a number larger than 4. **19**	Use squares and triangles to make an "AB" pattern. **20**	Write the names for the numbers: 0, 1, 2, 3, 4, and 5. Circle the number words with the fewest letters. **21**
Write the names for the numbers: 6, 7, 8, 9, and 10. Circle the number words with the most letters. **22**	Buddy put his 5 stuffed animals in a line. The bear is not first nor is it last. It is in an odd numbered place. In which place is the bear? **23**	How many different outfits can be made using red, orange, and yellow shirts and green, blue, and purple shorts? Draw a picture showing the different outfits. **24**

Math Practice Assignment Sheet Set #2

Name: ______________________________ **Due Date:** ________________

Directions: During this month complete at least 18 of the questions. After completing the questions, have your teacher or parent initial the work sheets. Some of the questions will need to be completed on a separate sheet of paper. Staple those pages to the back of the work sheets. Return the work sheets and all attached pages on the assigned due date.

Draw a picture of an object that is taller than you and a picture of something that is shorter than you. **1**	Take 6 beans. Use a marker to color one side of each bean black. Shake the beans and pour them on the table. How many beans landed black side up? Write down the math problem. Do this 6 times. **2**	Draw a picture and write a math problem. A monkey eats a banana each day. How many bananas can a monkey eat in one week? **3**
Draw a picture and write a math problem. A cat and a dog walked in the garden. How many paw prints would they leave behind? **4**	Draw a picture and write a math problem. When Dinah was born, she was 2 feet long. Now she is 4 feet long. How many feet has Dinah grown? **5**	Dudley went bowling. On his turn he knocked down 10 pins. On his next turn, Dudley knocked down 4 pins. How many pins did Dudley knock down in all? **6**
Draw a picture and write a math problem. A ladybug has 3 spots on its left side. It has the same number of spots on the right side. How many spots are on 2 ladybugs? **7**	Draw a picture and write a math problem. There were 15 scouts in the forest. Then 3 of the scouts got lost. How many scouts did not get lost? **8**	Write the numbers 0 to 10. Circle all the numbers with digits or numbers bigger than 3. **9**
Make an "ABC" pattern using squares, circles, and triangles. **10**	On a piece of paper, design a flag. Write a sentence describing your flag. **11**	Brainstorm about fish. Make a list of all the different kinds of fish that you can think of. **12**

Math Practice Assignment Sheet Set #2 *(cont.)*

Name: ______________________________ **Due Date:** ________________

Jack can buy 3 magic beans for 5¢ or 8 magic beans for 10¢. Which one is the better deal? Why?

13

Draw a picture showing three different ways that 18¢ can be made. Use "p" for penny, "n" for nickel, or "d" for dime.

14

Write the missing numbers.

12, ___, 14, ___, 16, 17

4, ___, ___, ___, 8

10, ___, ___, ___, 14

15

Write the number that comes next.

9, ______ 4, ______

0, ______ 7, ______

16

Write the numbers 0 to 10. Circle the numbers that are smaller than 4. Draw a square around the numbers that are larger than 4.

17

Complete each math problem.

3 + ______ = 3

6 + ______ = 10

12 + ______ = 15

10 + ______ = 12

18

Draw a picture and write a math problem. Only 6 kids can fit on a school bus. How many school buses are needed to take 12 students on a field trip?

19

Write the number that comes before.

______,11

______,10

______, 8

______, 4

20

Draw a picture and write a math problem. Only 8 apples can fit into one basket. Mary picked 10 apples. How many baskets does Mary need?

21

Make tally marks to show each number.

7 ______________

2 ______________

9 ______________

5 ______________

22

When Tiffany went for a walk, she saw 6 cats, 5 birds, and 4 ants. How many bugs did Tiffany see in all?

23

Draw a picture for this math problem. Which has more wheels—a tricycle or a 2–wheel bike?

24

Math Practice Assignment Sheet Set #3

Name: ______________________ **Due Date:** ______________

Directions: During this month complete at least 18 of the questions. After completing the questions, have your teacher or parent initial the work sheets. Some of the questions will need to be completed on a separate sheet of paper. Staple those pages to the back of the work sheets. Return the work sheets and all attached pages on the assigned due date.

Fact Family: Use the numbers 2, 4, and 6 to make 2 addition problems and 2 subtraction problems. **1**	Mystery Number: I am a number larger than 5 but less than 10. I am an even number. I am greater than 6. What number am I? **2**	Draw a picture and write a math problem. Henrietta Hamster loves to eat lettuce. Each day she eats 2 pieces of lettuce. How many pieces of lettuce does Henrietta eat in one week? **3**
Put the numbers in order, smallest to greatest. 7, 5, 12, 9, 13 ____ ____ ____ ____ ____ **4**	Write the numbers 0 to 20. How many 2's did you use? **5**	Add. 12 + 5 = ______ 3 + 14 = ______ 11 + 6 = ______ 7 + 10 = ______ **6**
Subtract. 10 – 5 = ______ 7 – 4 = ______ 18 – 6 = ______ 19 – 10 = ______ **7**	Fill 2 measuring cups (a ½ cup and ¼ cup) with water. Which measuring cup holds more water? **8**	Which would you rather have—½ of a pie or ¼ of a pie? Why? **9**
How many days are in this month? Write down all of the dates (numbers) in this month. **10**	Circle all of the numbers that have a 3 in them. 7 13 10 23 3 16 **11**	Use a measuring tape to measure your hand and your foot. Which one is longer? **12**

Math Practice Assignment Sheet Set #3 *(cont.)*

Name: ______________________________ **Due Date:** ____________________

Play tic-tac-toe. How many "x's" were used in the game? How many "o's" were used in the game? **13**	Mental Math 3 + 2 + 1 − 4 = ____ 8 − 4 + 2 + 0 = ____ 7 − 7 + 5 − 4 = ____ 10 − 9 + 3 − 0 = ____ **14**	Complete each math problem. 13 + ____ = 16 14 + ____ = 17 19 + ____ = 19 15 + ____ = 18 **15**
Margaret sells a glass of lemonade for 25¢. One afternoon she sold 4 lemonades. How much money did Margaret make? **16**	Complete each math problem. 17 − ____ = 14 15 − ____ = 5 11 − ____ = 10 16 − ____ = 12 **17**	Write the numbers 0 to 10. Circle all of the even numbers. **18**
Eric loves to collect baseball cards! On Monday he bought 3 cards. On Tuesday, he bought 8 more. How many cards did Eric buy in all? **19**	Write 5 addition and/or subtraction problems that have 9 as the answer. **20**	Fact Family: Use the numbers 1, 9, and 10 to make 2 addition problems and 2 subtraction problems. **21**
Write the numbers 11 to 20. Circle all of the odd numbers. **22**	Draw a picture and write a math problem. Jennifer walks 3 blocks to the store and then 6 more blocks to her friend's house. How many blocks does Jennifer walk in all? **23**	Wesley ran his first lap in 4 minutes. He ran a second lap in 2 minutes. His third lap was the total time of the first and second laps. How many minutes did Wesley run in all? **24**

Math Practice Assignment Sheet Set #4

Name: ______________________________ **Due Date:** ________________

Directions: During this month complete at least 18 of the questions. After completing the questions, have your teacher or parent initial the work sheets. Some of the questions will need to be completed on a separate sheet of paper. Staple those pages to the back of the work sheets. Return the work sheets and all attached pages on the assigned due date.

Draw a picture and write a math problem. Jeffrey has 13 green caps and 4 yellow caps. How many caps does Jeffrey have in all? **1**	Draw a picture for this math problem. Juanita collects stamps. She has 13 stamps with hearts and stamps with flowers. She has 23 stamps in all. How many stamps have flowers on them? **2**	Add. 13 + 16 = ______ 14 + 13 = ______ 10 + 9 = ______ 12 + 7 = ______ **3**
Subtract. 10 – 5 = ______ 19 – 9 = ______ 16 – 4 = ______ 18 – 11 = ______ **4**	Draw a picture and write a math problem. Mr. Green gave Julio 6 packets of seeds. Each packet had 3 seeds in it. How many seeds did Mr. Green give Julio? **5**	Write the missing number. 24 + ______ = 28 24 + ______ = 25 12 + ______ = 18 11 + ______ = 19 **6**
Draw a picture and write a math problem. Yvette saw 2 birds. They both had the same number of feathers. Yvette counted 18 feathers in all. How many feathers did each bird have? **7**	Draw a picture and write a math problem. Garrett had 12 video games. He bought 2 more games. How many games does Garrett have in all? **8**	Mystery Number: I am greater than 6 and less than 16. I am an odd number. When you count by 5's, you say my name. What number am I? **9**
Write the number that comes next. 23, ______ 14, ______ 30, ______ 49, ______ **10**	Fact Family: Use the numbers 5, 8, and 13 to make 2 addition problems and 2 subtraction problems. **11**	Draw a picture and write a math problem. If everybody in your class brought 1 penny to school, how much money would there be in all? **12**

Math Practice Assignment Sheet Set #4 *(cont.)*

Name: ______________________ **Due Date:** ______________

Write the number that comes before.

_____, 100
_____, 81
_____, 17
_____, 15

13

Write the missing number.

29 – _____ = 23
24 – _____ = 20
19 – _____ = 6
18 – _____ = 7

14

Ask your family if they like pizza or spaghetti better. Make a chart showing how each family member voted. Write a sentence about your findings.

15

Write your phone number. Circle all of the odd digits.

16

Make tally marks to show each number.

20
12
14
9

17

Who is the tallest person in your family? Who is the shortest person in your family?

18

Draw a picture and write a math problem. How many different ways could you put 9 stars into 2 jars?

19

Add and subtract.

12 + 4 – 3 + 11 = _____
19 – 6 + 2 – 5 = _____
20 – 5 + 0 – 3 = _____
18 – 3 + 5 – 11 = _____

20

Draw a picture and write a math problem. If everybody in your family bought a pair of roller skates, how many pairs would they need to buy? How many single roller skates would there be in all?

21

Look at each pair of numbers. Circle the smaller number.

20 15
8 13
16 10

22

Make a list showing all of the birthdays in your family. Put the dates (numbers) in order from smallest to largest.

23

Write the number that comes in the middle.

45, _______, 47
59, _______, 61
82, _______, 84

24

Math Practice Assignment Sheet Set #5

Name: ______________________________ **Due Date:** ______________

Directions: During this month complete at least 18 of the questions. After completing the questions, have your teacher or parent initial the work sheets. Some of the questions will need to be completed on a separate sheet of paper. Staple those pages to the back of the work sheets. Return the work sheets and all attached pages on the assigned due date.

Draw a picture and write a math problem. Cut an apple in half and take out the seeds. How many seeds are in the apple? Use the seeds to make a math problem. **1**	Draw a picture and write a math problem. Marco saw some snakes. He saw 10 eyes in all. How many snakes did he see? **2**	At the soda shop, a soda costs 60¢. Each refill costs 35¢. Louise drank 2 glasses of soda. How much money did Louise spend? **3**
Draw a picture and write a math problem. When emptying his pockets, Felipe found 5 nickels and 3 pennies. How much money did Felipe find in all? **4**	Write the missing number. 13 + ______ = 15 4 + ______ = 12 7 + ______ = 14 10 + ______ = 11 **5**	Draw a picture and write a math problem. There are 9 people at the beach. Then 3 go into the water. How many people are on the sand? **6**
Draw a picture and write a math problem. Erica has 2 quarters, 1 dime, and 3 pennies. Does she have enough money to buy the 70¢ ball? **7**	Write the number words for the numbers 11, 12, 13, 14, and 15. Circle the longest number word. **8**	Make a graph and write a sentence about the following information. Six friends like pumpkin pie. Four friends like apple pie. Five friends like cherry pie. **9**
Draw a picture and write a math problem. There are 10 people standing in line. Jody is in the 3rd space. How many people are ahead of Jody? How many people are behind Jody? **10**	Write the number that is 6 more. 10 ______ 4 ______ 1 ______ 12 ______ **11**	Write the missing number. 15 – ______ = 9 14 – ______ = 14 12 – ______ = 3 15 – ______ = 7 **12**

Math Practice Assignment Sheet Set #5 *(cont.)*

Name: ______________________ **Due Date:** ______________

Draw a picture and write a math problem. A pie can be cut into 8 slices. How many pies are needed for 14 people? **13**	Write the number words for the numbers 16, 17, 18, 19, and 20. Circle the shortest number word. **14**	Hannah was standing in line at the movies. There were 4 people ahead of her and 6 people behind her. How many people were standing in the line? **15**
Which is worth more: 2 quarters, 4 dimes, or 8 nickels? What is the total value of each set of coins? **16**	Draw a picture and write a math problem. When Francisco went to the beach, he saw 6 starfish, 4 seashells, and 5 beach balls. How many sea creatures did Francisco see? **17**	Add. 16 + 3 = ______ 14 + 2 = ______ 12 + 2 = ______ 24 + 1 = ______ **18**
Draw a picture and write a math problem. How many different ways can 12 pieces of candy be put into 3 candy jars? **19**	Make a design or picture using 12 toothpicks. Glue the toothpicks to a piece of paper. Write a sentence about what you made. **20**	Draw a picture and write a math problem. Ryan planted 14 pumpkin seeds. Only 5 of the seeds sprouted. How many seeds did not sprout? **21**
Subtract. 24 – 8 = ______ 22 – 9 = ______ 25 – 6 = ______ 23 – 8 = ______ **22**	Grace asked 15 people if they like carrots. Only 8 people said, "yes," the rest said "no." Make a graph and write a sentence about Grace's findings. **23**	Mystery Number: If you add 6 to the number then subtract 3 the answer is 12. What is the mystery number? **24**

Math Practice Assignment Sheet Set #6

Name: ______________________________ **Due Date:** ________________

Directions: During this month complete at least 18 of the questions. After completing the questions, have your teacher or parent initial the work sheets. Some of the questions will need to be completed on a separate sheet of paper. Staple those pages to the back of the work sheets. Return the work sheets and all attached pages on the assigned due date.

1
Celeste has 1 quarter, 3 dimes, 4 nickels, and 12 pennies. How much money does Celeste have in all?

2
Make 4 addition and subtraction problems that have 15 as the answer.

3
Look at a hundreds board. Write the numbers that are 27's "neighbors." (Neighbors are the numbers that are above, below, in front of, and in back of 27.)

4
Write numbers to make each statement true.

_______ < _______

_______ > _______

_______ < _______

_______ > _______

5
Look at a hundreds board. Write down the numbers that have a 4 in the "ones" place.

6
Write the missing numbers.

25, ___, 27, ___, ___, 30, ___, ___, ___,34, ___, ___, ___, 38, ___, ___

7
Write the number that comes between.

98, _______, 100

77, _______, 79

63, _______, 65

8
Draw a picture and write a math problem. Nathan has 18 balloons. Keeping none for himself, he gives each one of his friends 3 balloons. How many people did Nathan share balloons with?

9
Write the number that is 8 more.

10 _______

7 _______

12 _______

3 _______

10
Tyler's favorite number is 3. Zachary's number is twice as big as Tyler's plus 5 more. What is Zachary's number?

11
Fact Family: Use the numbers 8, 9, and 17 to make 2 addition problems and 2 subtraction problems.

12
Write the number that is 10 less.

15 _______

20 _______

19 _______

10 _______

Math Practice Assignment Sheet Set #6 *(cont.)*

Name: ______________________ **Due Date:** ______________

Subtract,

12 – 7 = ________

22 – 8 = ________

24 – 4 = ________

18 – 5 = ________

13

What is the largest number you can make using the digits 0 and 8? What is the smallest number you can make using the digits 0 and 8?

14

Draw a picture and write a math problem. Hector has a quarter, 2 dimes, and 5 pennies. How much money does Hector have?

15

Add.

27 + 8 = ________

29 + 7 = ________

10 + 6 = ________

12 + 6 = ________

16

Write your favorite 1-digit number. Write your favorite 2-digit number.

17

Write the numbers from 1 to 100. Circle all the numbers used when counting by 10's.

18

Count by 3's to 30. Write the numbers on a piece of paper.

19

How many pennies are needed to make the following:

1 nickel ________
1 dime ________
1 quarter ________
1 half dollar ________
1 dollar ________

20

Draw a picture and write a math problem. Rosa won 12 tickets at the carnival. Moses won 5 fewer tickets than Rosa. How many tickets did Moses win?

21

Write the value for the following coins: penny, nickel, dime, quarter, and half dollar.

22

Draw a picture and write a math problem. There are 16 marbles. Each child has 4 marbles. How many children are there?

23

Draw a picture and write a math problem. Miguel bought 3 movie tickets for $7 each. How much money did Miguel spend?

24

Math Practice Assignment Sheet Set #7

Name: ______________________ **Due Date:** ______________

Directions: During this month complete at least 18 of the questions. After completing the questions, have your teacher or parent initial the work sheets. Some of the questions will need to be completed on a separate sheet of paper. Staple those pages to the back of the work sheets. Return the work sheets and all attached pages on the assigned due date.

Draw a picture and write a math problem. Maurice has 16 rocks. Only 9 of them are round. The rest of them are flat. How many flat rocks does Maurice have? **1**	Write the numbers 0 to 20. Circle all of the odd numbers. How many odd numbers are there? How many even numbers are there? **2**	Draw a picture and write a math problem. Arden can do 5 math problems in 10 minutes. How long will it take Arden to do 15 math problems? **3**
Draw a picture and write a math problem. Mrs. Davis needs 3 eggs to make a cake. How many eggs does Mrs. Davis need in order to make 6 cakes? **4**	On a clock, there are 2 hands. What are the 2 hands called? What do they show on the clock? **5**	How many minutes are in a quarter (1/4) of an hour? How many minutes are in half (1/2) an hour? **6**
Fact Family: Use the numbers 7, 9, and 16 to make 2 addition problems and 2 subtraction problems. **7**	What is another way of writing 12:30? **8**	How many minutes are in one hour? In 12 hours? In 24 hours? **9**
Look at each pair of numbers. Circle the smaller number. 68 12 89 100 45 14 **10**	Draw a picture and write a math problem. Earl earns 12¢ for each paper he sells. How many papers does Earl need to sell to make 48¢? **11**	Look at each pair of numbers. Circle the larger number. 25 52 18 9 90 78 **12**

Math Practice Assignment Sheet Set #7 *(cont.)*

Name: ______________________________ **Due Date:** ________________

Keep track of how you spend your time. Pick one day. Record what time you get up, what you do every hour, and what time you go to bed. **13**	Keep track of time! How many minutes does it take you to get to school each morning? **14**	Draw a picture and write a math problem. The birthday party started at 2:00. Adrian arrived 15 minutes late. What time did Adrian arrive at the party? **15**
Draw a picture and write a math problem. Each big truck has 6 tires. How many tires are on 5 trucks? **16**	Subtract. 19 – 11 = ________ 32 – 12 = ________ 31 – 10 = ________ 28 – 9 = ________ **17**	Draw a picture and write a math problem. The octopus caught 22 sea creatures: 6 starfish, 5 seahorses, 7 jellyfish, and some clams. How many clams did the octopus catch? **18**
Look at a clock. Write down the numbers that are on the face of the clock. **19**	Make up 5 addition and/or subtraction problems with 16 as the answer. **20**	Look in this month's calendar. How many days are there between the first day of the month and the last day of the month? (Don't count the first and last days.) **21**
Make tally marks to show each number. 25 17 **22**	Draw a picture and write a math problem. The 4 Robinson children are 4 years apart in age. The youngest child is 3 years old. How old are the other 3 children? **23**	Mystery Number: I am greater than 25 and less than 50. When you count by 9's you say my name. When you count by 12's you say my name. What number am I? **24**

Math Practice Assignment Sheet Set #8

Name: ______________________________ **Due Date:** ________________

Directions: During this month complete at least 18 of the questions. After completing the questions, have your teacher or parent initial the work sheets. Some of the questions will need to be completed on a separate sheet of paper. Staple those pages to the back of the work sheets. Return the work sheets and all attached pages on the assigned due date.

Draw a picture and write a math problem. A pack of gum costs 60¢. If Theresa has 75¢, how much change would she have after buying a pack of gum? **1**	Count by 2's to 100. Write the numbers down on a piece of paper. Circle all of the numbers that have a 6 in the ones place. **2**	Add. 25 + 52 = ______ 60 + 7 = ______ 81 + 11 = ______ 36 + 40 = ______ **3**
Take out a deck of cards and remove all of the face cards. Add up the numbers on all of the remaining cards. What is the total? **4**	Cross out the number that is not needed. 87 21 – 21 0 **5**	The Perez sisters went to a baseball game. The total cost for the tickets was $85. If there are 5 sisters in the Perez family, what was the ticket price for each sister? **6**
Subtract. 87 – 57 = ______ 91 – 90 = ______ 26 – 15 = ______ 44 – 22 = ______ **7**	Cross out the number that is not needed. 27 35 + 22 49 **8**	Count by 5's to 100. Write the numbers down on a piece of paper. Circle all of the numbers that have a 5 in the ones place. **9**
If Amir were given 25¢ a day for 30 days, how much money would Amir have? **10**	Count by 10's to 100. Write the numbers down on a piece of paper. Circle all of the numbers that have a 0 in the ones place. **11**	It is 15 miles from Mary's house to Grandma's house. It is 22 miles from Grandma's house to Mary's aunt's house. How many miles from Mary's house to Mary's aunt's house? **12**

Math Practice Assignment Sheet Set #8 *(cont.)*

Name: ______________________________ **Due Date:** ________________

If Jane was given a dime and 2 nickels every day, how much money would Jane have in 7 days? **13**	Draw a picture and write a math problem. One spider has 2 eyes and 8 legs. How many eyes and legs do 3 spiders have? **14**	Write how many tens and ones for each number. 54 _____tens _____ ones 72 _____tens _____ ones 81 _____tens _____ ones **15**
Draw a picture and write a math problem. One jellyfish has 7 stingers. There are 35 stingers. How many jelly fish are there? **16**	How many triangles do you see below? **17**	Draw a picture and write a math problem. There are 15 students in Esmeralda's class. Esmeralda finished her test 5th. How many students finished before her? How many students finished after her? **18**
Draw a picture and write a math problem. To cut a pizza pie into 12 equal slices, how many straight cuts would you need to make through the center of the pie? **19**	Draw a picture and write a math problem. Birthday candles cost 6¢ each. How much would it cost to put birthday candles on your birthday cake for this year? **20**	Look on a hundreds board. Write down the numbers that are "neighbors" to 93. **21**
How many ways can 25¢ be made? Use "q" for quarter, "d" for dime, "n" for nickel, and "p" for penny to show the different ways to make 25¢. **22**	Samantha has 17 pencils. How many ways can she put the 17 pencils into two pencil cups? Make a list showing the different combinations. **23**	How many different numbers under 400 can be made using the numbers 1, 2, and 3 one time each in a 3-digit number? **24**

Math Practice Assignment Sheet Set #9

Name: ______________________________ **Due Date:** ______________

Directions: During this month complete at least 18 of the questions. After completing the questions, have your teacher or parent initial the work sheets. Some of the questions will need to be completed on a separate sheet of paper. Staple those pages to the back of the work sheets. Return the work sheets and all attached pages on the assigned due date.

Write the number that is 10 more.

25 _____

80 _____

90 _____

61 _____

[1]

Write the missing numbers.

81, ___, ___, 84, ___, ___, 87, ___, ___, ___, 91, ___, ___, ___, ___, ___, 97, ___, ___

[2]

How many different ways can you count to 100? Write down the different ways of counting to 100.

[3]

Make a list of things that always come in pairs. (Examples: twins, shoes, eyes, etc.)

[4]

How many quarters does it take to make $5?

[5]

Andy earns $1 for mowing lawns and 25¢ for cleaning a window. Andy mowed 5 lawns and washed 3 windows. How much money did Andy earn?

[6]

Add.

27 + 41 = _______

55 + 44 = _______

91 + 3 = _______

80 + 7 = _______

[7]

Make 5 addition and/or subtraction problems that have 19 as the answer.

[8]

Write the missing numbers.

50, 49, ___, ___, ___, 45, ___, ___, ___, ___, 40, ___, ___, ___, 36, ___, ___, ___, ___, 31

[9]

Write the number that is 10 less.

83 _______

41 _______

12 _______

55 _______

[10]

Draw a picture and write a math problem. Barney spent 50¢ on candy and $1.50 on games. How much did Barney spend?

[11]

Write numbers to make each statement true.

_______ < _______

_______ > _______

_______ < _______

_______ > _______

[12]

Math Practice Assignment Sheet Set #9 *(cont.)*

Name: ______________________________ **Due Date:** ____________________

13 Count by 25's to 100. Write the numbers down. Circle all of the even numbers.

14 Subtract.

22 – 10 = ________

31 – 30 = ________

59 – 9 = ________

40 – 40 = ________

15 Count by 2's from 100 to 200. Write the numbers. Circle the numbers that have a 2 in the ones place.

16 Write down three different ways to make 80¢. Use "q" for quarter, "d" for dime, "n" for nickel, and "p" for penny.

17 Draw a picture and write a math problem. Darla had 87 flowers. She sold 43 flowers for $1 each. How many flowers are left? How much money did she make?

18 Mystery Number: I am more than 25 and less than 100. When you count by 25, you say my name. When you count by 10, you say my name. What number am I?

19 It costs Bridget 45¢ to make each cupcake. She sells them for 65¢ each. How much money does Bridget make on each cupcake she sells?

20 Grandma gave Floyd 1 quarter, 4 dimes, 1 nickel, and 8 pennies. How much money did Grandma give to Floyd?

21 Nicholas has 5 coins in his wallet. Together they make 38¢. Three of the coins are the same. What coins does Nicholas have in his wallet?

22 Each necklace has 22 beads and 10 teddy bears. How many items are on one necklace?

23 Each vowel (a, e, i, o, u) is worth 10¢. Each consonant is worth 5¢. How many words can you make that are worth exactly 25¢? 50¢?

24 Sarah has 4 coins in her pocket. Together they total 17¢. What coins are in Sarah's pocket?

Math Journal Review Question Set #1

Name: ______________________________ **Date:** ____________________

1. Write the numbers 0 to 10.

____ ____ ____ ____ ____ ____ ____ ____ ____ ____ ____

2. Count the number of stars in each box. Write the number on the line.

☆ ☆ ☆ ☆ ☆ ☆ ☆ ☆ | ☆ ☆ ☆ ☆ ☆ | ☆ | ☆ ☆ | ☆ ☆ ☆ ☆ ☆ ☆ ☆ ☆ ☆ | ☆ ☆ ☆

________ ________ ________ ________ ________ ________

3. Write the addition sentence.

\+

____________ + ____________ = ____________

4. Write the subtraction sentence.

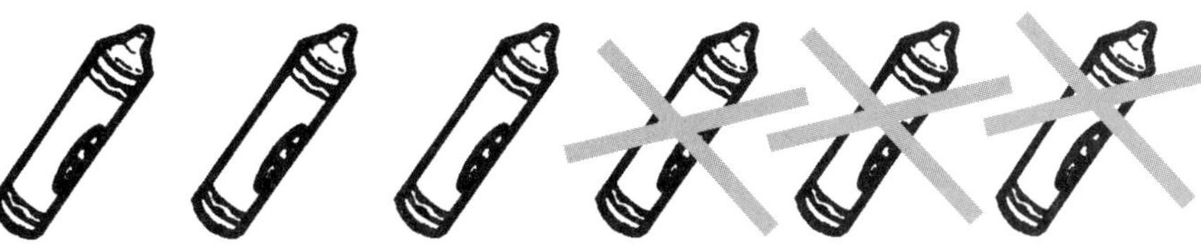

____________ – ____________ = ____________

Math Journal Review Question Set #1 *(cont.)*

Look at the Venn diagram. Answer the questions.

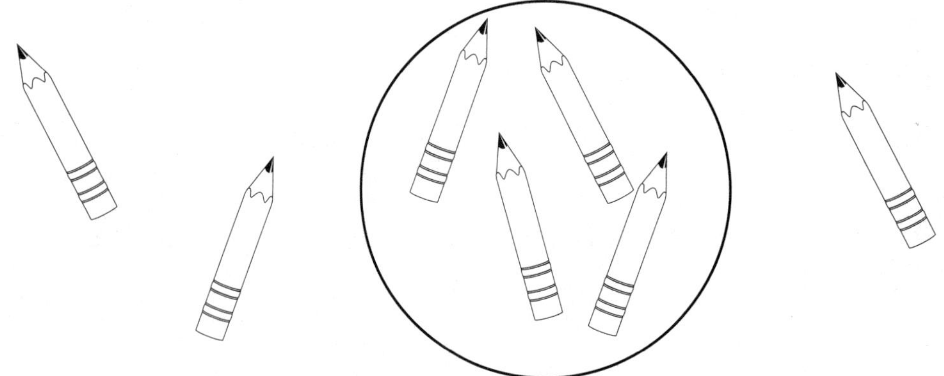

5. How many pencils are in the circle? ____________________
6. How many pencils are not in the circle? ____________________
7. How many pencils are there in all? ____________________
8. Complete the pattern.

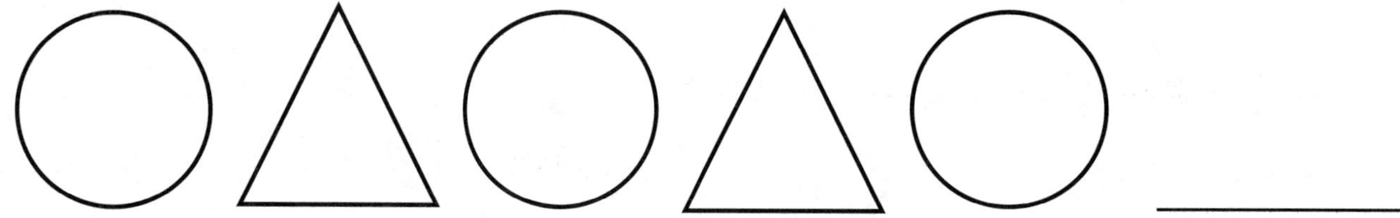

9. Add.

1 + 1 = ______ 2 + 3 = ______ 4 + 1 = ______

10. Subtract.

3 – 0 = ______ 5 – 2 = ______ 2 – 2 = ______

Math Journal Review Question Set #2

1. Draw a line to match each number word to its number.

ten	0
one	1
three	2
eight	3
nine	4
zero	5
four	6
two	7
seven	8
five	9
six	10

2. Write the number that comes before.

_______, 4 _______, 10 _______, 7

3. Write the number that comes after.

2, _______ 8, _______ 5, _______

4. Write the number that comes in the middle.

0, _______, 2 3, _______, 5 7, _______, 9

Math Journal Review Question Set #2 *(cont.)*

5. Count the pennies below. Write the amount of money on the line.

________________ ________________ ________________

6. Use tally marks to show each number.

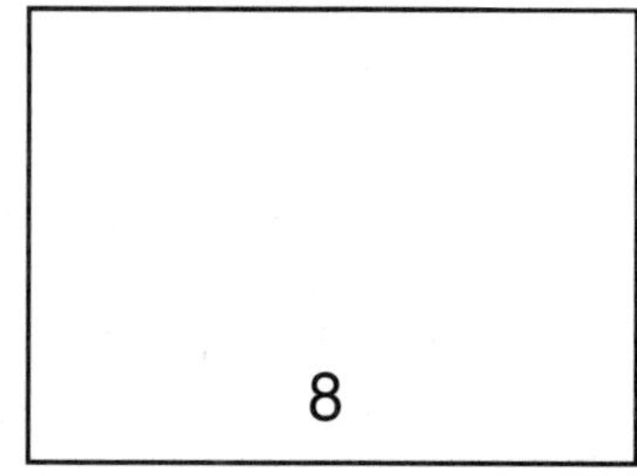

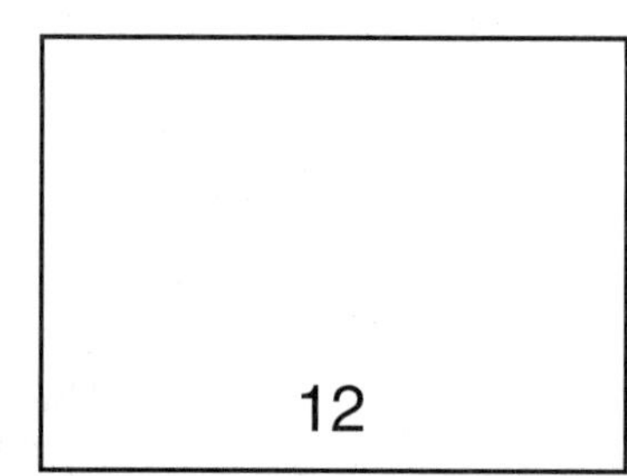

7. Fact Family—use the numbers 2, 3, and 5 to make 2 addition problems and 2 subtraction problems.

_______ + _______ = 5 5 – _______ = _______

_______ + _______ = 5 5 – _______ = _______

8. Add.

5 + 1 = ______ 3 + 2 = ______ 1 + 0 = ______

9. Subtract.

4 – 2 = ______ 6 – 0 = ______ 3 – 1 = ______

Math Journal Review
Question Set #3

Name: ______________________________ **Date:** ____________________

Look at the Venn diagram. Answer the questions.

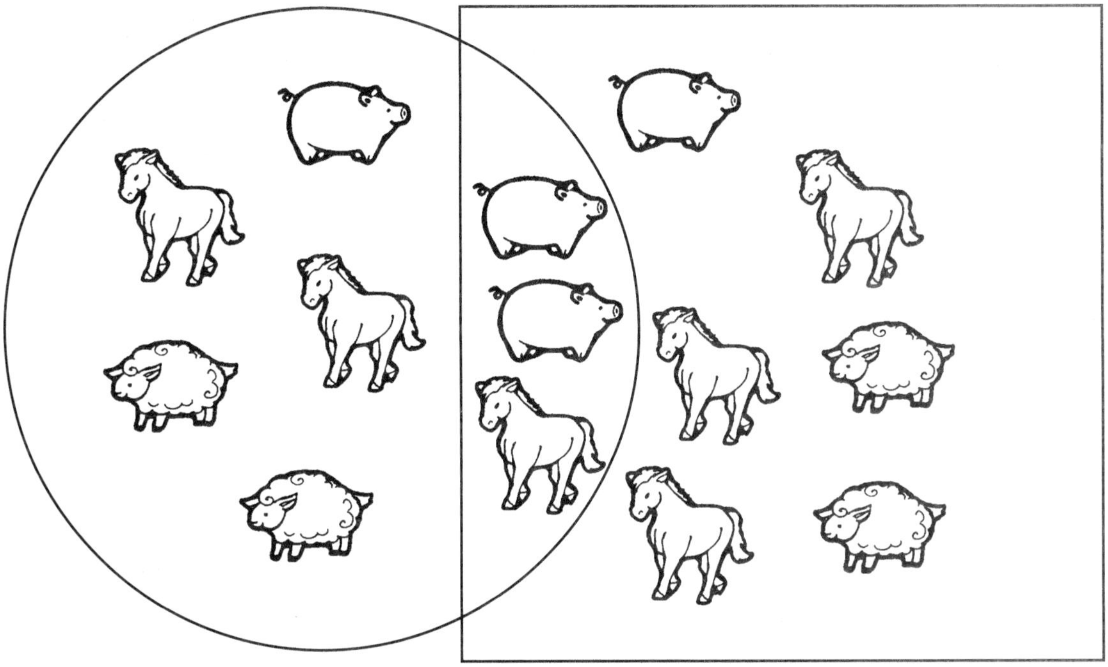

1. How many pigs are in the square? __________

2. How many sheep are in the circle? __________

3. How many horses are in the square, but not in the circle? __________

4. Which farm animal is not in both the square and the circle? __________

5. Use the numbers 3, 5, and 8 to make two addition problems and 2 subtraction problems.

_______ + _______ = _______   _______ – _______ = _______

_______ + _______ = _______   _______ – _______ = _______

Math Journal Review Question Set #3 *(cont.)*

6. Solve the word problem. Write a math problem and a sentence about the answer.

Christopher has this amount of change in his pocket. How much change does Christopher have in his pocket?

7. Count the number of corners on each shape. Write the number on the line.

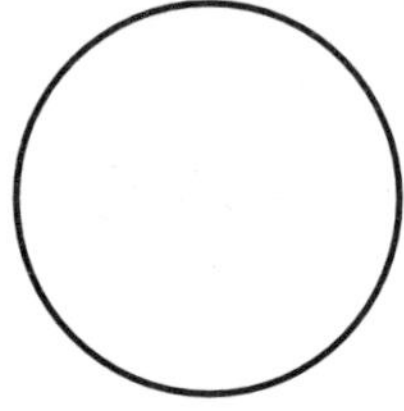 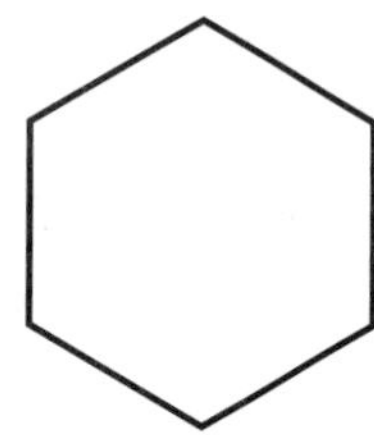 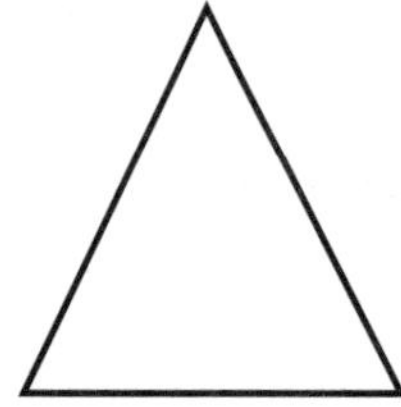 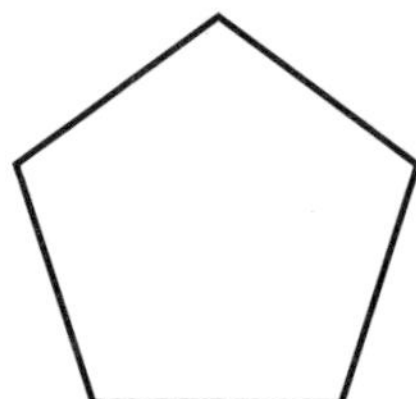 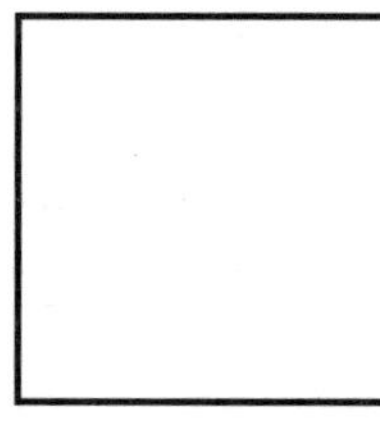

________ ________ ________ ________ ________

8. Add

13 + 4 = ________

2 + 16 = ________

11 + 5 = ________

5 + 10 = ________

9. subtract.

18 – 7 = ________

17 – 3 = ________

14 – 4 = ________

15 – 2 = ________

Math Journal Review Question Set #4

Name: ______________________________ **Date:** ____________________

1. Compare the 2 numbers. Use the > (greater than) and < (less than) symbols. Complete each sentence.

12 ◯ 18	15 ◯ 1	0 ◯ 11
____ is less than ____	____ is greater than ____	____ is less than ____

2. Add the 3 numbers.

3 + 2 + 4 = ________ 2 + 5 + 0 = ________ 3 + 2 + 1 = ________

3. Write the math sentence.

Joey saw 10 ladybugs. Three ladybugs flew away. How many ladybugs were left?

____

4. Write the math sentence.

Maggie saw 4 caterpillars. Then Maggie saw 5 more caterpillars. How many caterpillars did Maggie see in all?

____

5. Write the missing numbers.

11, ________,13 ________,17, 18 12, 13, ________

Math Journal Review Question Set #4 *(cont.)*

Look at the graph on ladybugs. Answer the questions.

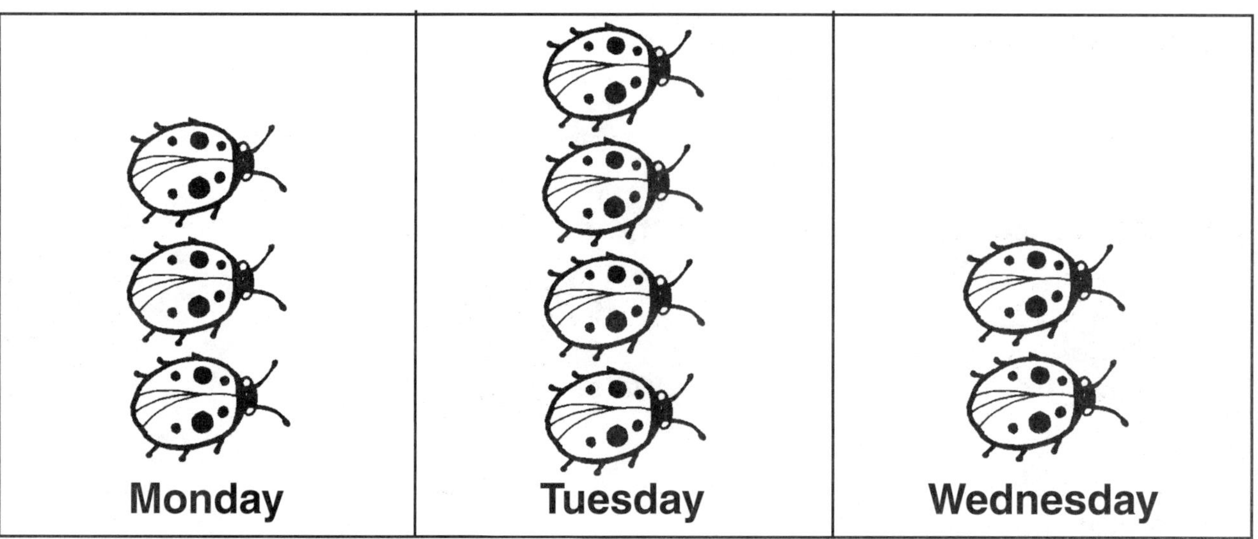

6. How many ladybugs were seen on Monday? ________________

7. How many ladybugs were seen on Tuesday? ________________

8. Were more ladybugs seen on Tuesday or Wednesday? ________________

9. How many ladybugs were seen in all? ________________

Add or subtract.

10. **25** + **14**

11. **28** − **12**

12. **17** + **32**

13. **19** − **10**

Math Journal Review Question Set #5

Name: ______________________ **Date:** ______________

1. Write the numbers 11 to 20.

____ ____ ____ ____ ____ ____ ____ ____ ____ ____

2.Count the money. Write the totals on the lines.

____

____

____

____

____

____

3. Add or subtract.

10 + 20 = ________

11 + 14 = ________

19 – 8 = ________

18 – 8 = ________

6 + 12 = ________

14 + 5 = ________

17 – 1 = ________

12 + 15 = ________

20 – 13 = ________

16 + 13 = ________

17 – 10 = ________

20 – 10 = ________

Math Journal Review Question Set #5 *(cont.)*

4. Look at the graph. Write a question about the graph.

Jason	△ △ △ △
Ashley	○ ○ ○ ○ ○ ○ ○

____

____

Look at the dogs. Follow the directions. (The dogs have been placed in order from left to right, starting at the top row.)

5. Put an X on the 1st dog.
6. Draw a O around the 3rd dog.
7. Draw a line under the 8th dog.
8. Draw a [ ] around the 5th dog.
9. Draw a triangle around the 2nd dog.
10. How many dogs were left alone? __________

Math Journal Review Question Set #6

Name: ______________________________ **Date:** ____________________

1. Write the number that is 3 more.

   1 ______ 12 ______ 10 ______ 16 ______ 1 ______

2. Write the number that is 3 less.

   5 ______ 9 ______ 17 ______ 8 ______ 3 ______

3. Match each number word to its number.

fourteen	11
twenty	12
sixteen	13
eighteen	14
twelve	15
fifteen	16
eleven	17
thirteen	18
seventeen	19
nineteen	20

4. Write a math sentence.

   Alex has a dozen eggs. He breaks 3 of the eggs. How many eggs are left?

   ____

Math Journal Review Question Set #6 *(cont.)*

Look at the objects below. Write one sentence about how they are alike. Write one sentence about how they are different.

5. Alike: ____________________

6. Different: ____________________

7. Write the missing number.

23 + ______ = 30	14 + ______ = 19	10 + ______ = 17
20 – ______ = 5	19 – ______ = 2	26 – ______ = 20

8. Use tally marks to show each number.

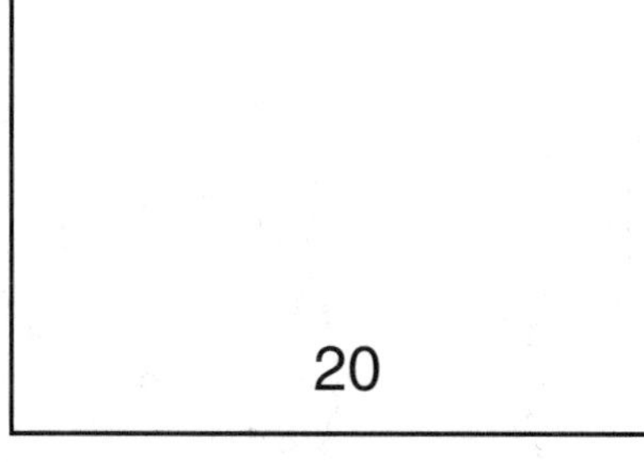
20

22

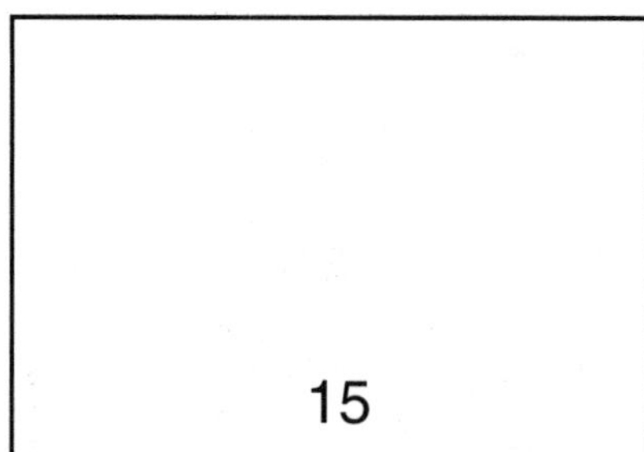
15

9. Use the numbers 6, 8, and 14 to make 2 addition problems and 2 subtraction problems.

______ + ______ = ______   ______ – ______ = ______

______ + ______ = ______   ______ – ______ = ______

Math Journal Review Question Set #7

Name: ______________________________ **Date:** ____________________

1. Write the time shown on each clock.

_____ : _____

_____ : _____

_____ : _____

_____ : _____

_____ : _____

_____ : _____

_____ : _____

_____ : _____

_____ : _____

_____ : _____

_____ : _____

_____ : _____

Math Journal Review Question Set #7 *(cont.)*

2. Add or subtract.

8 + 6 = ______	19 – 5 = ______	14 + 0 = ______	15 – 8 = ______
10 + 4 = ______	16 – 5 = ______	12 + 3 = ______	14 – 11 = ______
5 + 7 = ______	11 – 7 = ______	7 + 6 = ______	12 – 5 = ______

Solve each word problem by placing the correct math symbol in the box for each problem.

3. Joshua did 9 backflips. Then he did 4 more.
   How many backflips did Joshua do in all?

9

 4

4. Brandy did 15 somersaults. Her friend, Ashley did 9 fewer somersaults than Brandy did.
   How many somersaults did Ashley do?

15

 9

5. A tricycle has 3 wheels and 1 seat.
   How many wheels and seats are on 4 tricycles?

4

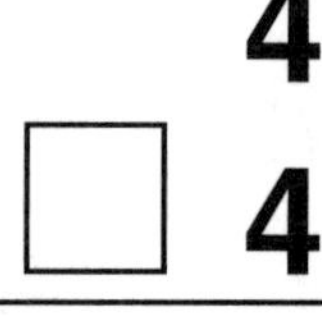 4

Math Journal Review Question Set #8

Name: ______________________________ **Date:** ____________________

1. Count by 2's to 20.

____ ____ ____ ____ ____ ____ ____ ____ ____ ____

2. Count by 5's to 50.

____ ____ ____ ____ ____ ____ ____ ____ ____ ____

3. Count by 10's to 100.

____ ____ ____ ____ ____ ____ ____ ____ ____ ____

4. Write the number of tens and ones.

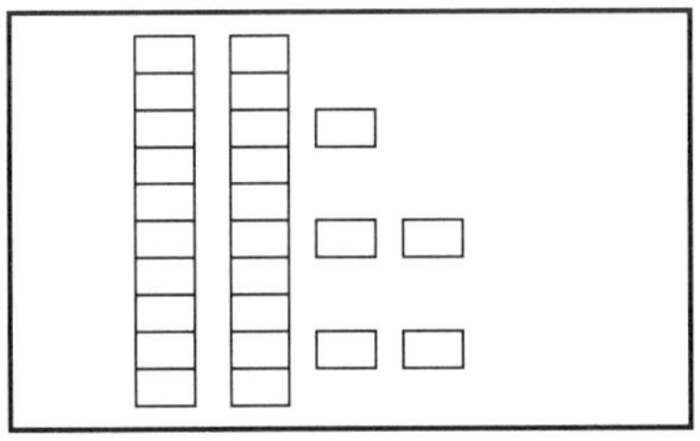

____ tens ____ ones 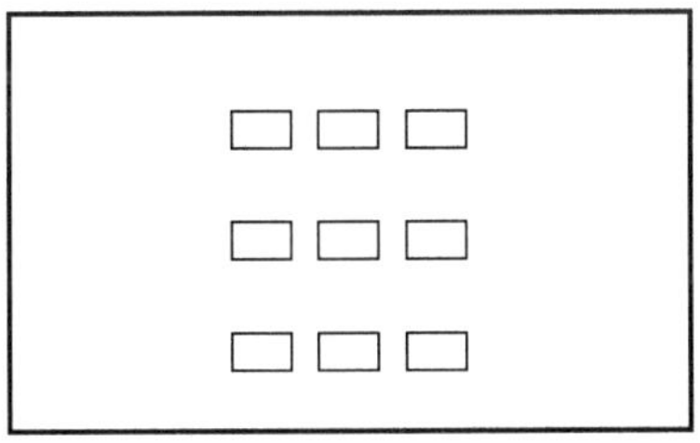 ____ tens ____ ones 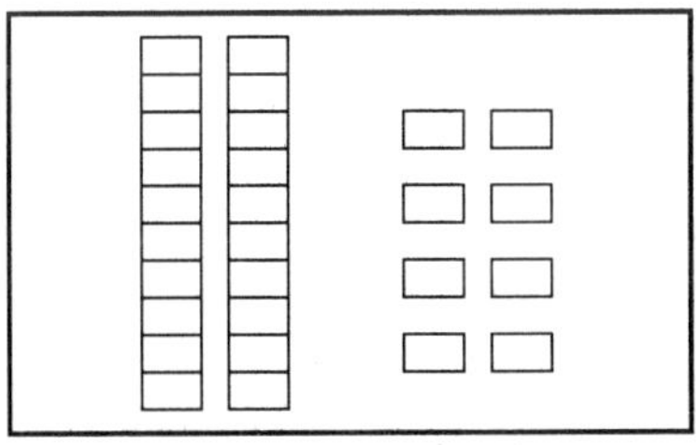 ____ tens ____ ones

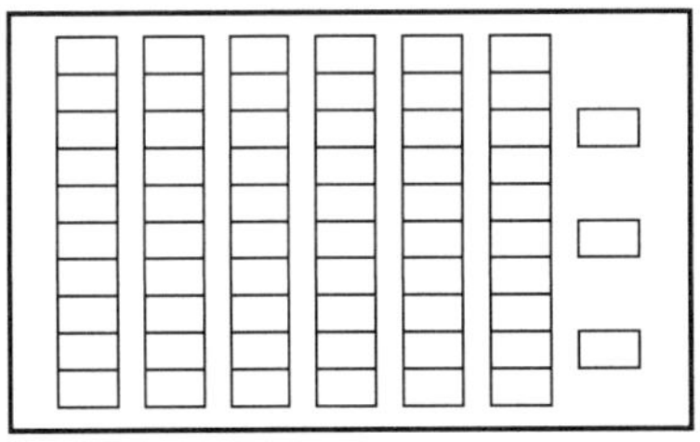

____ tens ____ ones 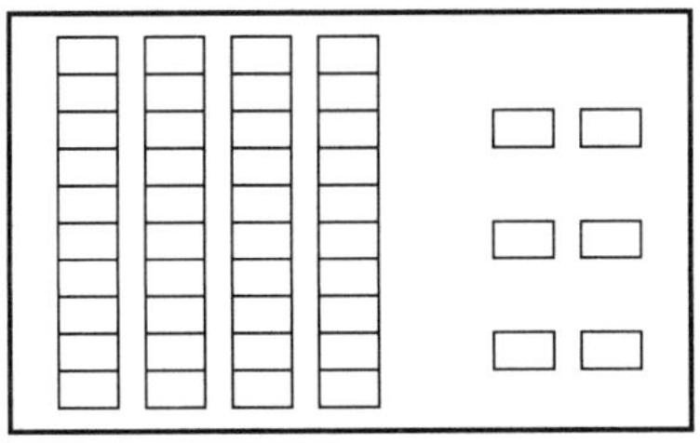 ____ tens ____ ones 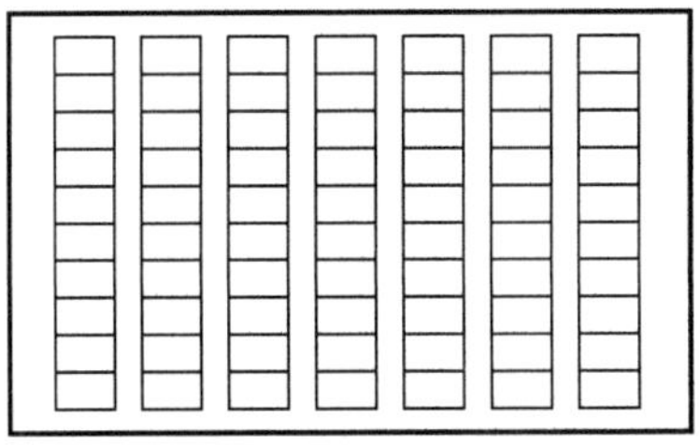 ____ tens ____ ones

Math Journal Review Question Set #8 *(cont.)*

5. Add and subtract.

22	44	57	46	65
+ 25	− 31	+ 42	+ 42	+ 3

80	19	99	45	77
− 70	+ 30	− 67	+ 40	− 7

Solve each word problem.

6. Seth had 25 marbles. ________

   His friend gave him 30 more marbles.

   How many marbles does Seth have now? + ________

   Seth has ____________ marbles in all.

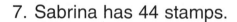

7. Sabrina has 44 stamps. ________

   She gives 22 stamps to her friend.

   How many stamps does Sabrina have left? − ________

   Sabrina has ____________ stamps left.

Math Journal Review Question Set #9

Name: ______________________ **Date:** ______________

Count the money. Write the total amount on the line.

1.

2.

3.

4.

5.
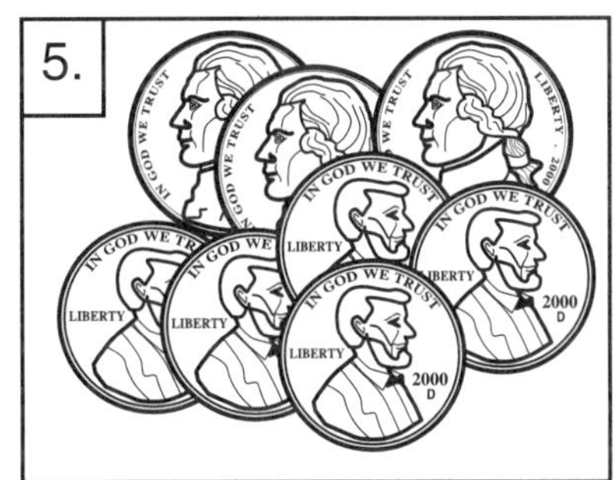

6.

7.
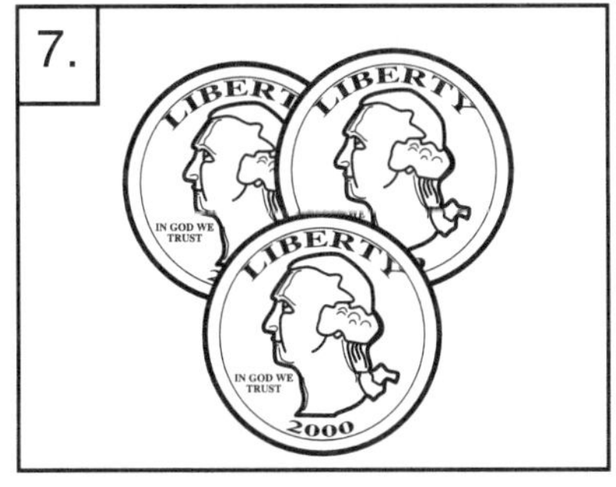

8.

9.
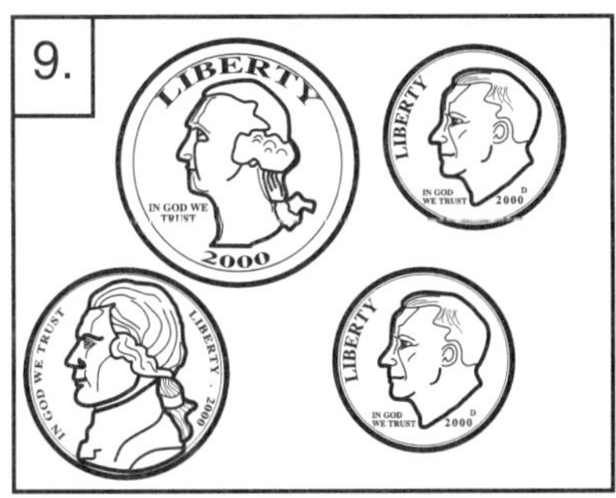

Draw the coins needed to show each amount. Use "q" for quarter, "d" for dime, "n" for nickel, and "p" for penny.

10.

37¢

11.

68¢

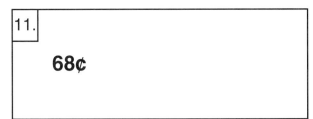

Math Journal Review Question Set #9 *(cont.)*

12. Calvin has $5.00 to spend at the bookstore. Make a list of the different combinations of items that Calvin can buy.

bookmark
$1.25

magazine
$2.75

newspaper
$.50

puzzles and games book
$3.45

coloring book
$3.25

small wall calendar
$4.45

notepad
$2.85

bookplate
$3.75

blank journal
$4.95

Answer Key

Teacher Note: *This answer key is for the Math Practice Assignment Sheet Sets and the Math Journal Review Question Sets in the Teacher's Supplement.*

Pages 5 and 6

1. Answer will vary.
2. 3 crayons
3. 6 pumpkins
4. 4 erasers
5. 5 bears

6.–10. Answers will vary.

11. 0 legs
12. bee, grasshopper, ladybug
13. 7 happy faces
14. 10 sheep
15. 3 golf balls
16. 3 boys
17. 7

18.–20. Answers will vary.

21. one and two
22. seven and eight
23. third
24. 9

Pages 7 and 8

1.–2. Answers will vary.

3. 7 bananas
4. 8 paw prints
5. 2 feet
6. 14 pins
7. 12 spots
8. 12 scouts
9. 4, 5, 6, 7, 8, 9, 10

10.–12. Answers will vary.

13. 8 magic beans for 10¢
14. possible answers: 1d 1n 3p; 3n 3p; 18p
15. 13, 15; 5, 6, 7; 11, 12, 13
16. 10; 5; 1; 8
17. numbers with circles: 0, 1, 2, 3
    numbers with squares: 5, 6, 7, 8, 9, 10
18. 3 + 0 = 3; 6 + 4 = 10; 12 + 3 = 15; 10 + 2 = 12
19. 2 buses
20. 10; 9; 7; 3
21. 2 baskets
22. Show number of tally marks as indicated in the problem.
23. 4 bugs
24. 2-wheel bike

Pages 9 and 10

1. 2 + 4 = 6; 4 + 2 = 6; 6 – 4 = 2; 6 – 2 = 4
2. 8
3. 14 pieces of lettuce
4. 5, 7, 9, 12, 13
5. 3
6. All problems equal 17.
7. 5; 3; 12; 9
8. 1/2 cup
9. 1/2 of a pie
10. Answer will vary.
11. 3, 13, 23

12.–13. Answers will vary.

14. 2; 6; 1; 4
15. 3; 3; 0; 3
16. $1.00
17. 3; 10; 1; 4
18. 2, 4, 6, 8, 10
19. 11 cards
20. Answer will vary.
21. 1 + 9 = 10; 9 + 1 = 10; 10 – 1 = 9; 10 – 9 = 1
22. 11, 13, 15, 17, 19
23. 9 blocks
24. 12 minutes

Pages 11 and 12

1. 17 caps
2. 10 stamps
3. 29; 27; 19; 19
4. 5; 10; 12; 7
5. 18 seeds
6. 4; 1; 6; 8
7. 9 feathers
8. 14 games
9. 15
10. 24; 15; 31; 50
11. 5 + 8 = 13; 8 + 5 = 13; 13 – 8 = 5; 13 – 5 = 8
12. Answer will vary.
13. 99; 80; 16; 14
14. 6; 4; 13; 11

15.–16. Answers will vary.

17. Show number of tally marks as indicated in the problem.

18.–19. Answers will vary.

20. 24; 10; 12; 9
21. Answer will vary.
22. 15, 8, 10
23. Answer will vary.
24. 46; 60; 83

Pages 13 and 14

1. Answer will vary.
2. 5 snakes
3. 95¢
4. 28¢
5. 2; 8; 7; 1
6. 6 people
7. No, Erica only has 63¢.
8. thirteen fourteen
9. Students may make varying types of graphs. Accept all possible answers.
10. 2 people in front, 7 people behind
11. 16; 10; 7; 18
12. 6; 0; 9; 8
13. 2 pies
14. twenty
15. 11 people
16. 2 quarters; 50¢, 40¢, 40¢
17. 6
18. 19; 16; 14; 25

19.–20. Answers will vary.

21. 9 seeds
22. 16; 13; 19; 15
23. Eight people like carrots and seven don't like carrots.
24. 9

Pages 15 and 16

1. 87¢
2. Answer will vary.
3. 17, 26, 28, 37
4. Answer will vary.
5. 4, 14, 24, 34, 54, 64, 74, 84, 94
6. 26, 28, 29, 31, 32, 33, 35, 36, 37, 39, 40
7. 99; 78; 64
8. 5 other people
9. 18; 15; 20; 11
10. 11
11. 8 + 9 = 17; 9 + 8 = 17; 17 – 9 = 8; 17 – 8 = 9
12. 5; 10; 9; 0
13. 5; 14; 20; 13
14. largest: 80; smallest: 08
15. 50¢
16. 35; 36; 16; 18
17. Answer will vary.
18. 10, 20, 30, 40, 50, 60, 70, 80, 90, 100
19. 3, 6, 9, 12, 15, 18, 21, 24, 27, 30
20. 5 pennies; 10 pennies; 25 pennies; 50 pennies; 100 pennies
21. 7 tickets
22. penny: 1 ; nickel: 5¢; dime: 10¢; quarter: 25¢; half dollar: 50¢
23. 4 children
24. $21

Pages 17 and 18

1. 7 rocks
2. 1, 3, 5, 7, 9, 11, 13, 15, 17, and 19 should be circled. There are 10 odd numbers, and 10 even numbers.
3. 30 minutes
4. 18 eggs
5. hour and minute hands; hour hand shows the hour; minute hand shows how many minutes
6. 60 minutes; 30 minutes
7. 7 + 9 = 16; 9 + 7 = 16; 16 – 7 = 9; 16 – 9 = 7
8. half past twelve
9. 15 minutes; 720 minutes; 1440 minutes
10. 12; 89; 14
11. 4 papers
12. 52; 18; 90

13.–14. Answers will vary.

15. 2:15
16. 30 tires
17. 8; 20; 21; 19
18. 4 clams

19.–21. Answers will vary.

22. Show number of tally marks as indicated in the problem.
23. 7, 11, 15
24. 36

Answer Key *(cont.)*

Pages 19 and 20

1. 15¢
2. 6, 16, 26, 36, 46, 56, 66, 76, 86, 96
3. 77; 67; 92; 76
4. Answer will vary.
5. 87 is not needed for the problem.
6. $17
7. 30; 1; 11; 22
8. 35 is not needed for the problem.
9. 5, 15, 25, 35, 45, 55, 65, 75, 85, 95
10. $7.50
11. 10, 20, 30, 40, 50 ,60, 70, 80, 90, 100
12. 37 miles
13. $1.40
14. 6 eyes and 24 legs
15. 5 tens 4 ones; 1 tens 2 ones; 8 tens 5 ones
16. 5 jellyfish
17. 17 triangles
18. Four students finished before Esmeralda, and 10 students finished after her.
19. 6 cuts
20. Answer will vary.
21. 83, 92, 94
22. Some possible answers: 1q; 2d 1n; 2d 5p; 1d 3n; 1d 2n 5p; 4n 5p; 5n; 25p

23.–24. Answers will vary.

Pages 21 and 22

1. 35; 90; 100; 71
2. 82, 83, 85, 86, 88, 89, 90, 92, 93, 94, 95, 96, 98, 99

3.–4. Answers will vary.

5. 20 quarters
6. $5.75
7. 68; 99; 94; 87
8. Answer will vary.
9. 48, 47, 46, 44, 43, 42, 41, 39, 38, 37, 35, 34, 33, 32
10. 72; 31; 2; 45
11. $2.00
12. Answer will vary.
13. 50, 100
14. 12; 1; 50; 0
15. 102, 112, 122, 132, 142, 152, 162, 172, 182, 192
16. Some possible answers: 3q 1n; 8 d; 2q 6n; 1q 1n 5d
17. 44 flowers left; $43
18. 50
19. 20¢
20. 78¢
21. 1 quarter, 1 dime, 3 pennies
22. 32
23. Answer will vary.
24. 1d 1n 2p

Pages 23 and 24

1. 0, 1, 2, 3, 4, 5, 6, 7, 8, 9, 10
2. 8; 5; 1; 2; 9; 3
3. 2 + 3 = 5
4. 6 – 3 = 3
5. 4
6. 3
7. 7
8. △ triangle
9. 2; 5; 5
10. 3; 3; 0

Pages 25 and 26

1. ten 10; one 1; three 3; eight 8; nine 9; zero 0; four 4; two 2; seven 7; five 5; six 6
2. 3; 9; 6
3. 3; 9; 6
4. 1; 4; 8
5. 2¢ ; 5¢ ; 7¢
6. Show number of tally marks as indicated in the problem.
7. 2 + 3 = 5; 3 + 2 = 5; 5 – 3 = 2; 5 – 2 = 3
8. 6; 5; 1
9. 2; 6; 2

Pages 27 and 28

1. 3 pigs
2. 2 sheep
3. 3 horses
4. sheep
5. 3 + 5 = 8; 5 + 3 = 8; 8 – 3 = 5; 8 – 5 = 3
6. 34¢
7. 0; 6; 3; 5; 4
8. 17; 18; 16; 15
9. 11; 14; 10; 13

Pages 29 and 30

1. $12 < 18$ so 12 is less than 18; $15 > 1$ so 15 is greater than 1; $0 < 11$ so 0 is less than 11
2. 9; 7; 6
3. There were 7 ladybugs left.
4. Maggie saw 9 caterpillars.
5. 12; 16; 14
6. 3
7. 4
8. Tuesday
9. 9 ladybugs
10. 39 butterflies
11. 16 dragonflies
12. 49 bees
13. 9 ants

Pages 31 and 32

1. 11, 12, 13, 14, 15, 16, 17, 18, 19, 20
2. 5¢ ; 10¢ ; 6¢ ; 10¢ ; 14¢ ; 9¢
3. 30; 25; 11; 10; 18; 19; 16; 27; 7; 29; 7; 10
4. Answer will vary.

5.–9. See the chart below.

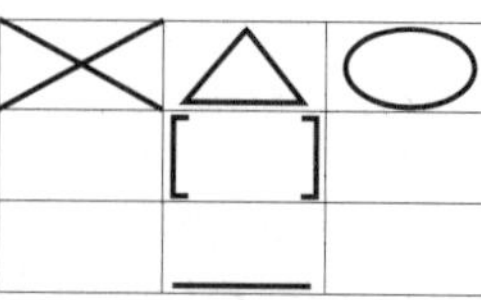

10. 4 dogs

Pages 33 and 34

1. 4; 15; 13; 19; 4
2. 2; 6; 14; 5; 0
3. fourteen 14; twenty 20; sixteen 16; eighteen 18; twelve 12; fifteen 15; eleven 11; thirteen 13; seventeen 17; nineteen 19
4. Alex has 9 eggs left.

5.–6. Answers will vary.

7. 7; 5; 7; 15; 17; 6
8. Show number of tally marks as indicated in the problem.
9. 6 + 8 = 14; 8 + 6 = 14; 14 – 6 = 8; 14 – 8 = 6

Pages 35 and 36

1. 4:00; 9:25; 2:00; 10:30; 3:25; 7:15; 12:45; 6:00; 4:55; 9:30; 10:30; 11:40
2. 14; 14; 14; 7; 14; 11; 15; 3; 12; 4; 13; 7
3. 9 + 4 = 13
4. 15 – 9 = 6
5. 4 x 4 = 16

Pages 37 and 38

1. 2, 4, 6, 8, 10, 12, 14, 16, 18, 20
2. 5, 10, 15, 20, 25, 30, 35, 40, 45, 50
3. 10, 20, 30, 40, 50, 60, 70, 80, 90, 100
4. 2 tens 5 ones; 0 tens 9 ones; 2 tens 8 ones; 6 tens 3 ones; 4 tens 6 ones; 7 tens 0 ones
5. 47; 13; 99; 88; 68; 10; 49; 32; 85; 70
6. 25 + 30 = 55: Seth has 55 marbles.
7. 44 – 22 = 22; Sabrina has 22 stamps left.

Pages 39 and 40

1. 40¢
2. 25¢
3. 50¢
4. 60¢
5. 20¢
6. 40¢
7. 75¢
8. 25¢
9. 50¢
10. Possible answers: 1q 1d 2p; 1q 2n 2p; 3d 1n 2p,
11. Possible answers: 2q 1d 8p; 6d 1n 3p; 5d 3n 3p
12. Possible answers: bookmark, magazine, and newspaper; coloring book, bookmark, and newspaper; blank journal

Numbers

Number Words

0 = zero
1 = one
2 = two
3 = three
4 = four
5 = five
6 = six
7 = seven
8 = eight
9 = nine
10 = ten
11 = eleven
12 = twelve
13 = thirteen
14 = fourteen
15 = fifteen
16 = sixteen
17 = seventeen
18 = eighteen
19 = nineteen
20 = twenty

Odd and Even

Odd numbers: 1, 3, 5, 7, 9

Even numbers: 0, 2, 4, 6, 8

Add +

together, in all, total, plus, add, more, greater

Subtract –

take away, subtract, minus, left, less, fewer

Skip Counting

Counting by 2's

0, 2, 4, 6, 8, 10, 12, 14, 16, 18, 20, 22, 24, 26, 28, 30, 32, 34, 36, 38, 40, 42, 44, 46, 48, 50

Counting by 5's

0, 5, 10, 15, 20, 25, 30, 35, 40, 45, 50, 55, 60, 65, 70, 75, 80, 85, 90, 95, 100

Counting by 10's

0, 10, 20, 30, 40, 50, 60, 70, 80, 90, 100

Money

half dollar
50¢

quarter
25¢

dime
10¢

nickel
5¢

penny
1¢

Addition Chart

+	0	1	2	3	4	5	6	7	8	9
0	0	1	2	3	4	5	6	7	8	9
1	1	2	3	4	5	6	7	8	9	10
2	2	3	4	5	6	7	8	9	10	11
3	3	4	5	6	7	8	9	10	11	12
4	4	5	6	7	8	9	10	11	12	13
5	5	6	7	8	9	10	11	12	13	14
6	6	7	8	9	10	11	12	13	14	15
7	7	8	9	10	11	12	13	14	15	16
8	8	9	10	11	12	13	14	15	16	17
9	9	10	11	12	13	14	15	16	17	18

Hundreds Chart

1	2	3	4	5	6	7	8	9	10
11	12	13	14	15	16	17	18	19	20
21	22	23	24	25	26	27	28	29	30
31	32	33	34	35	36	37	38	39	40
41	42	43	44	45	46	47	48	49	50
51	52	53	54	55	56	57	58	59	60
61	62	63	64	65	66	67	68	69	70
71	72	73	74	75	76	77	78	79	80
81	82	83	84	85	86	87	88	89	90
91	92	93	94	95	96	97	98	99	100

Rulers

Inch Rulers

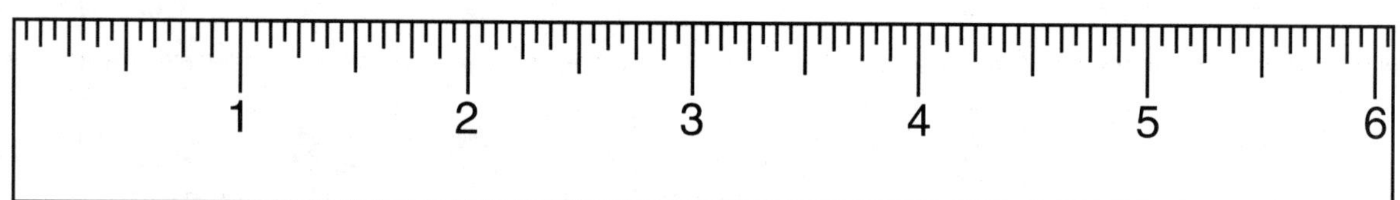

1 2 3 4 5 6

Centimeter Rulers

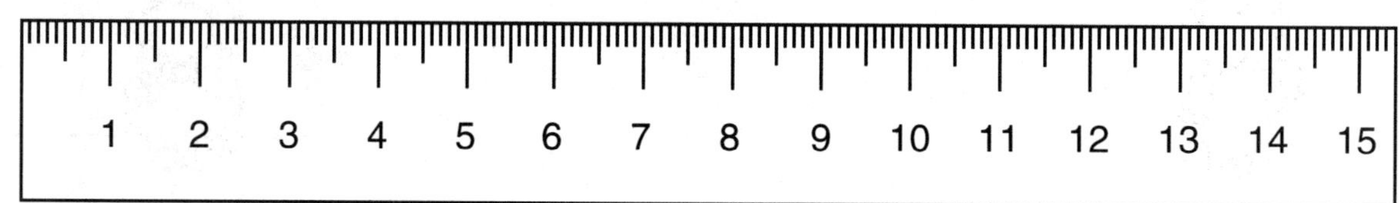

1 2 3 4 5 6 7 8 9 10 11 12 13 14 15

1 2 3 4 5 6 7 8 9 10 11 12 13 14 15

Number Cards

0	1	2
3	4	5
6	7	8
	9	

Number Lines

Photocopy the number lines onto cardstock or construction paper and cut apart. The students can store their number lines in the pocket inside their math journal.

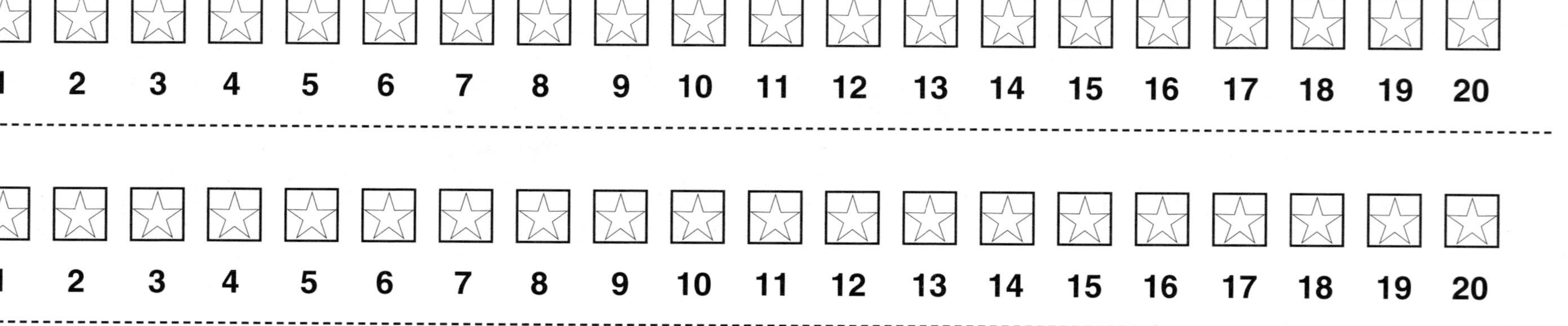

Number Lines

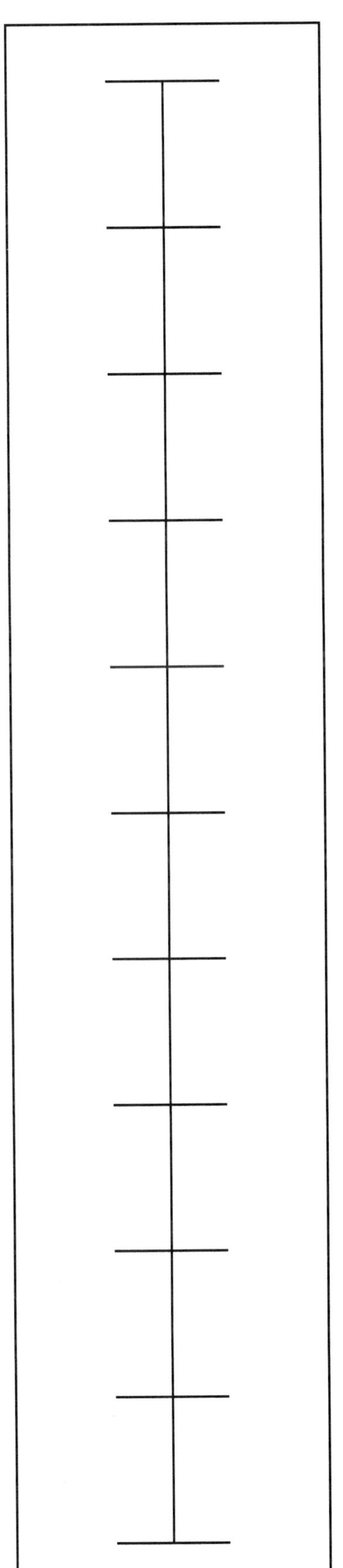

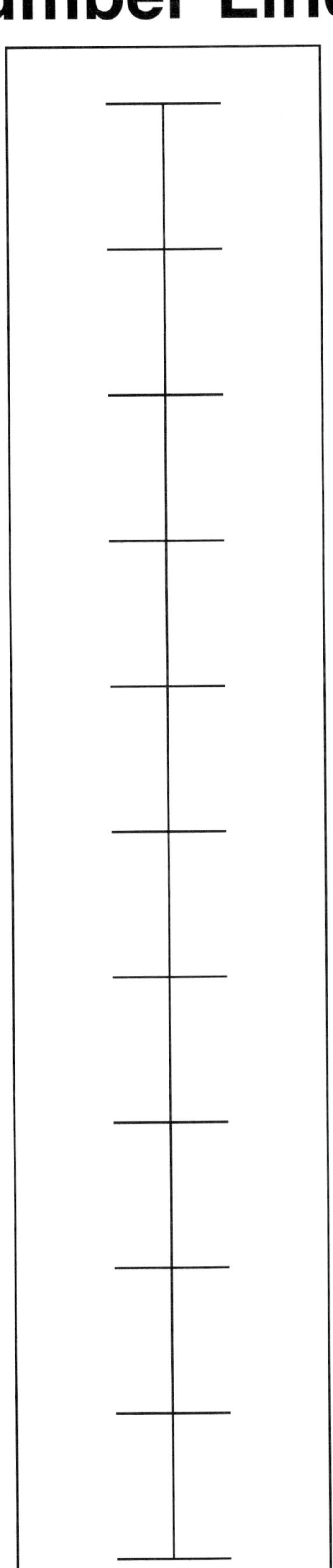

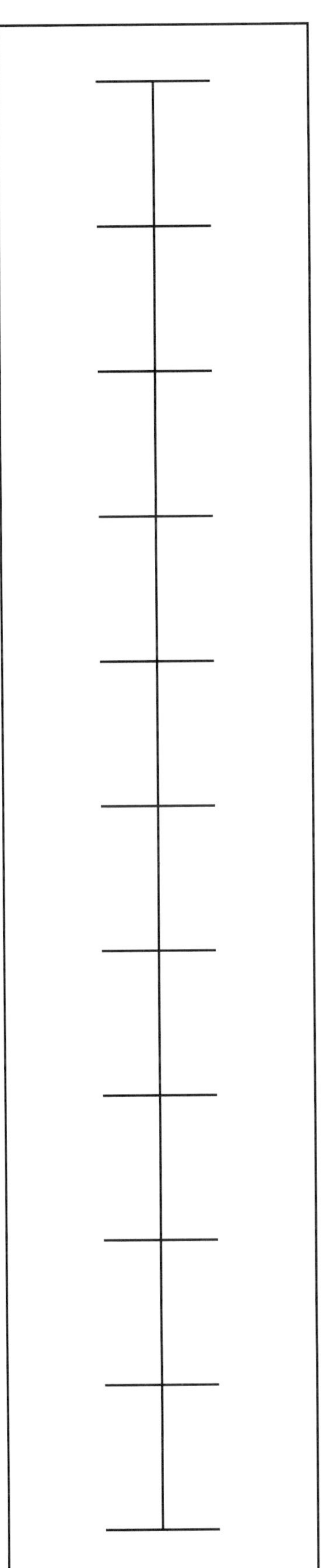

Place Value (Tens)

Place Value (Tens and Ones)

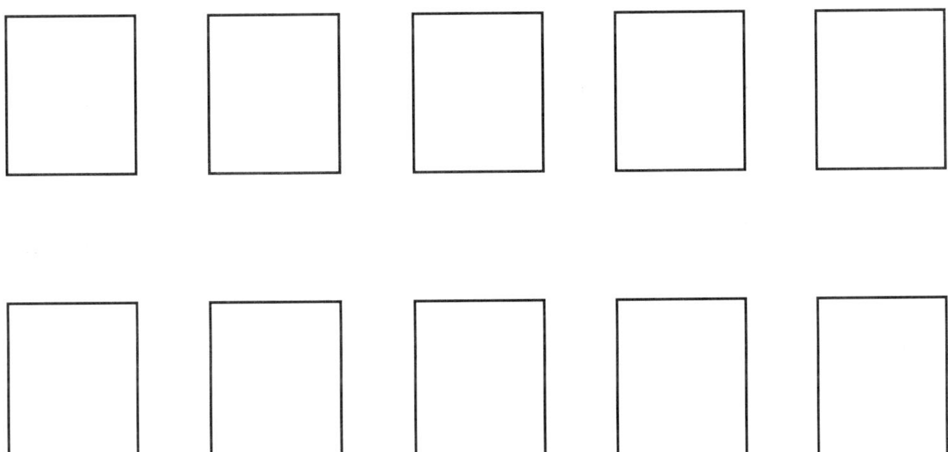

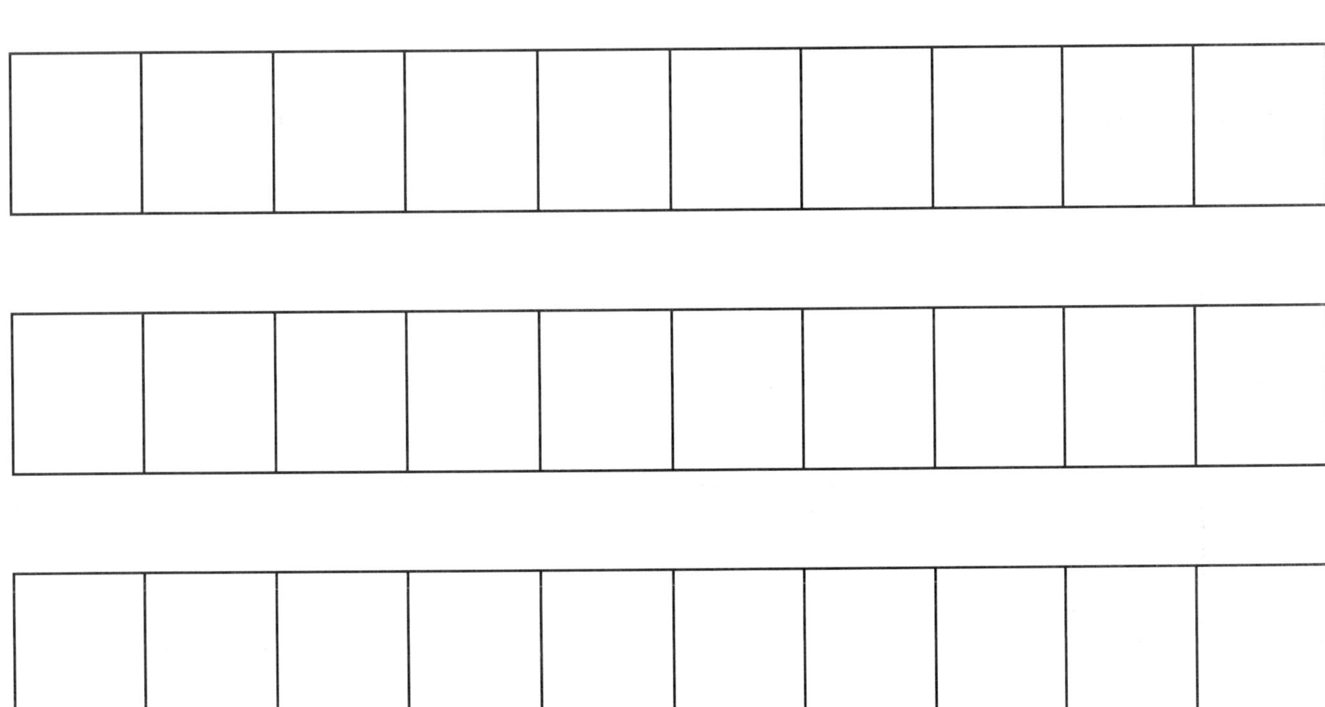

Place Value Chart

ones	
tens	

Place Value Mat

ones	
tens	
hundreds	

Math Symbols

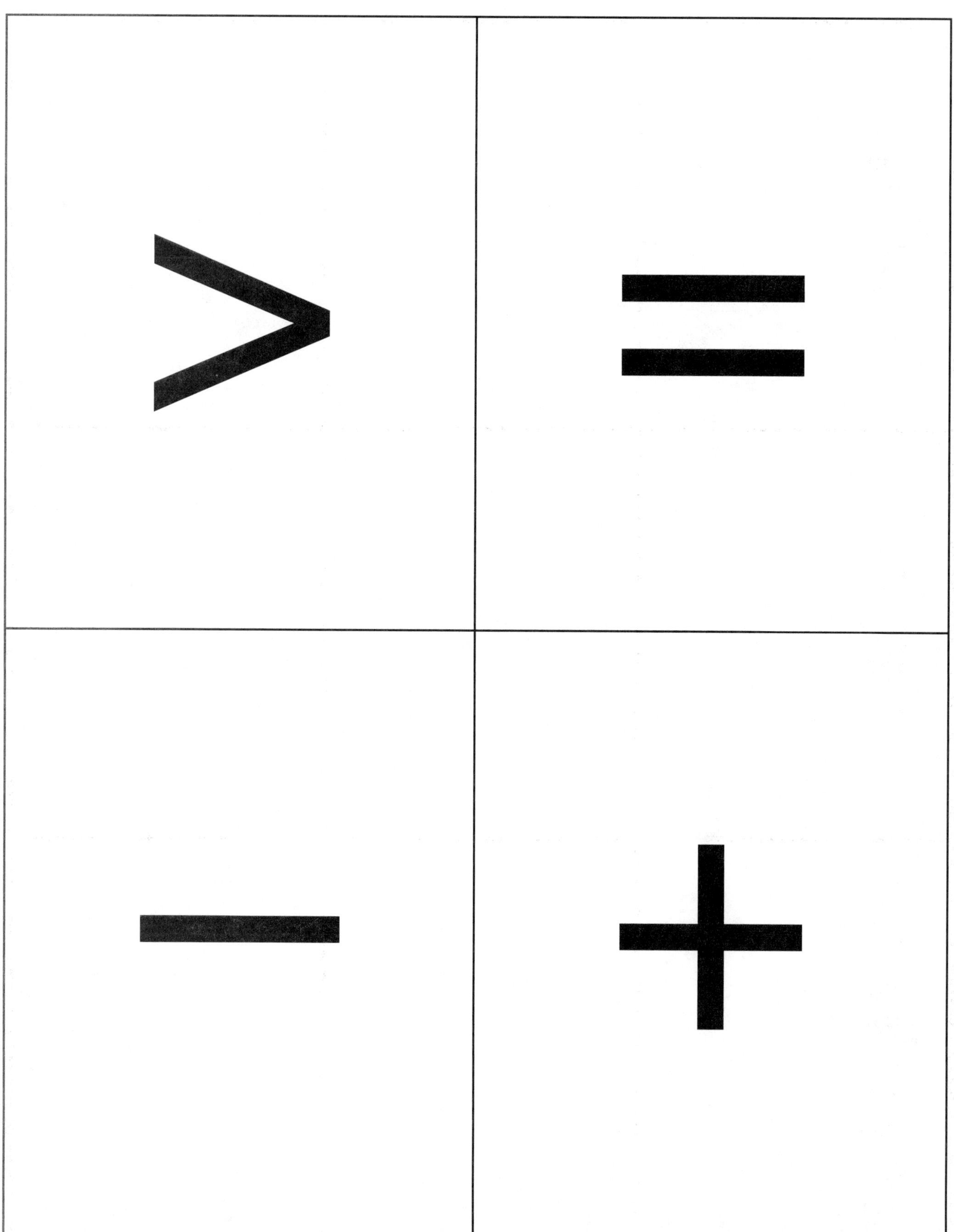

Number Puzzle Board

Money

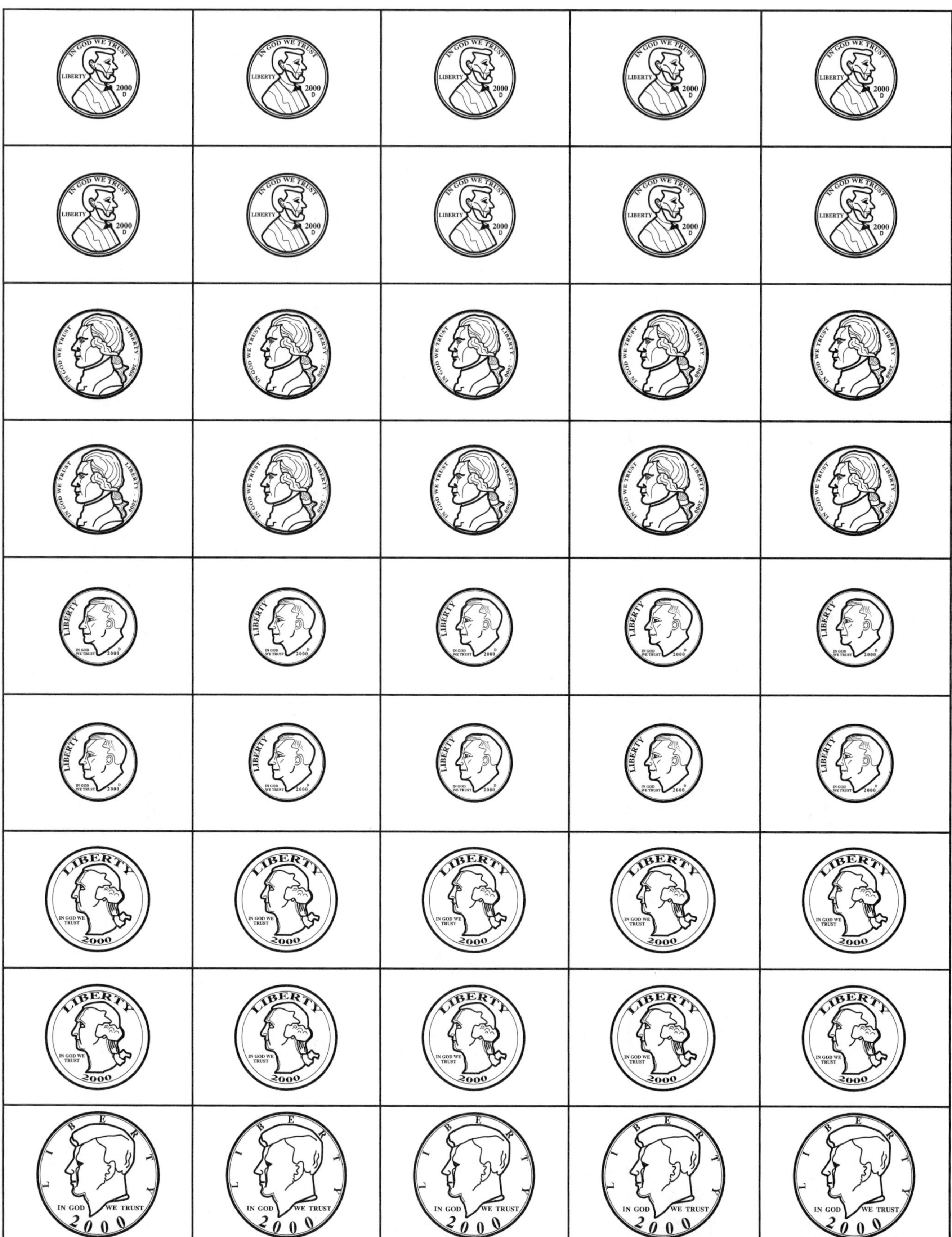

Clock Pattern

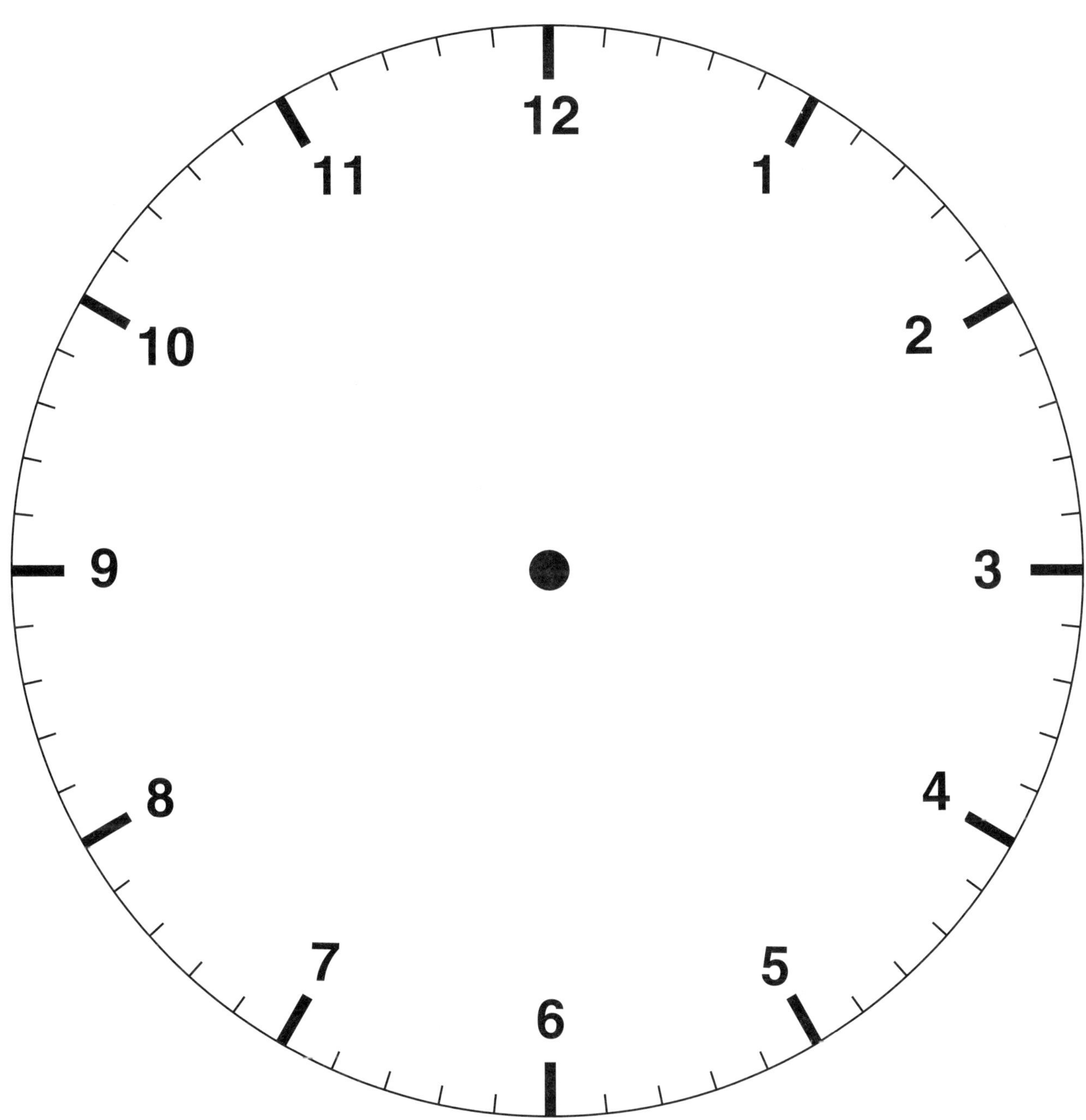

Shape Patterns

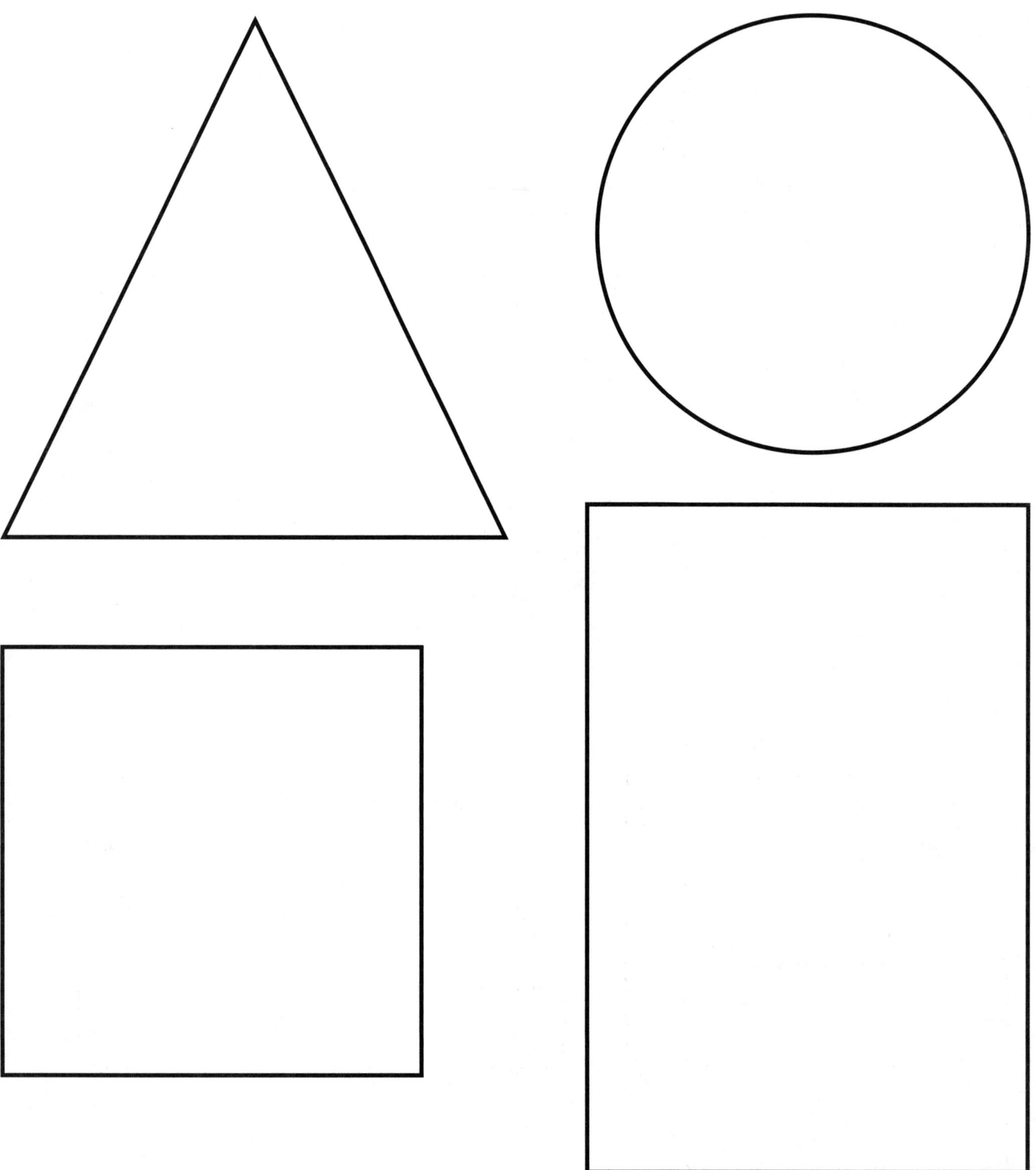

Shape Patterns *(cont.)*

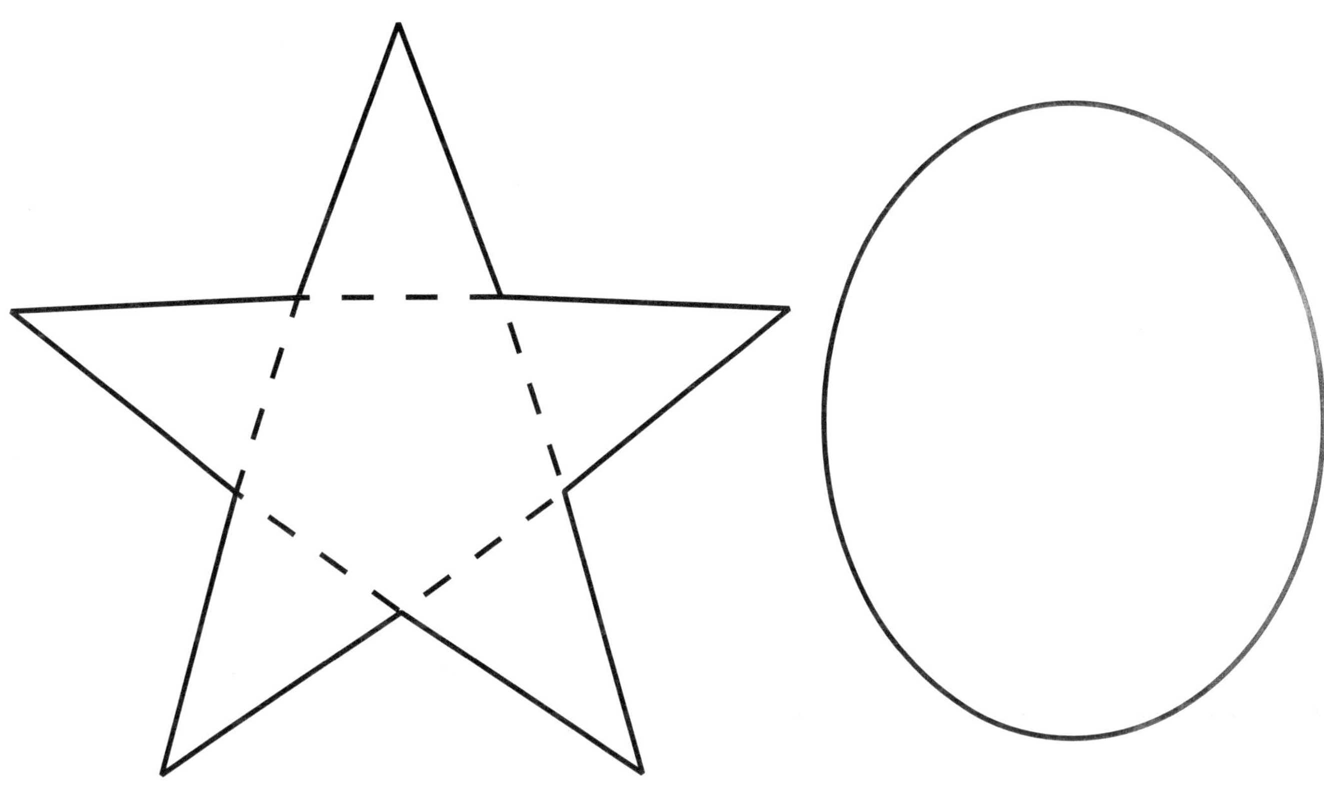

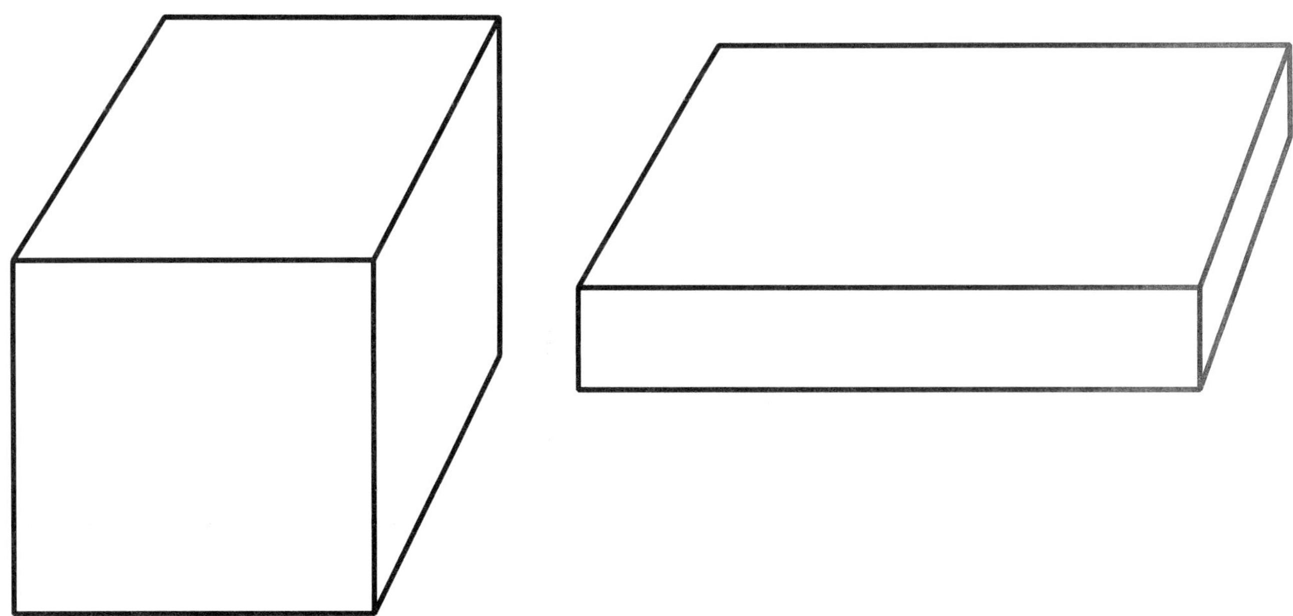

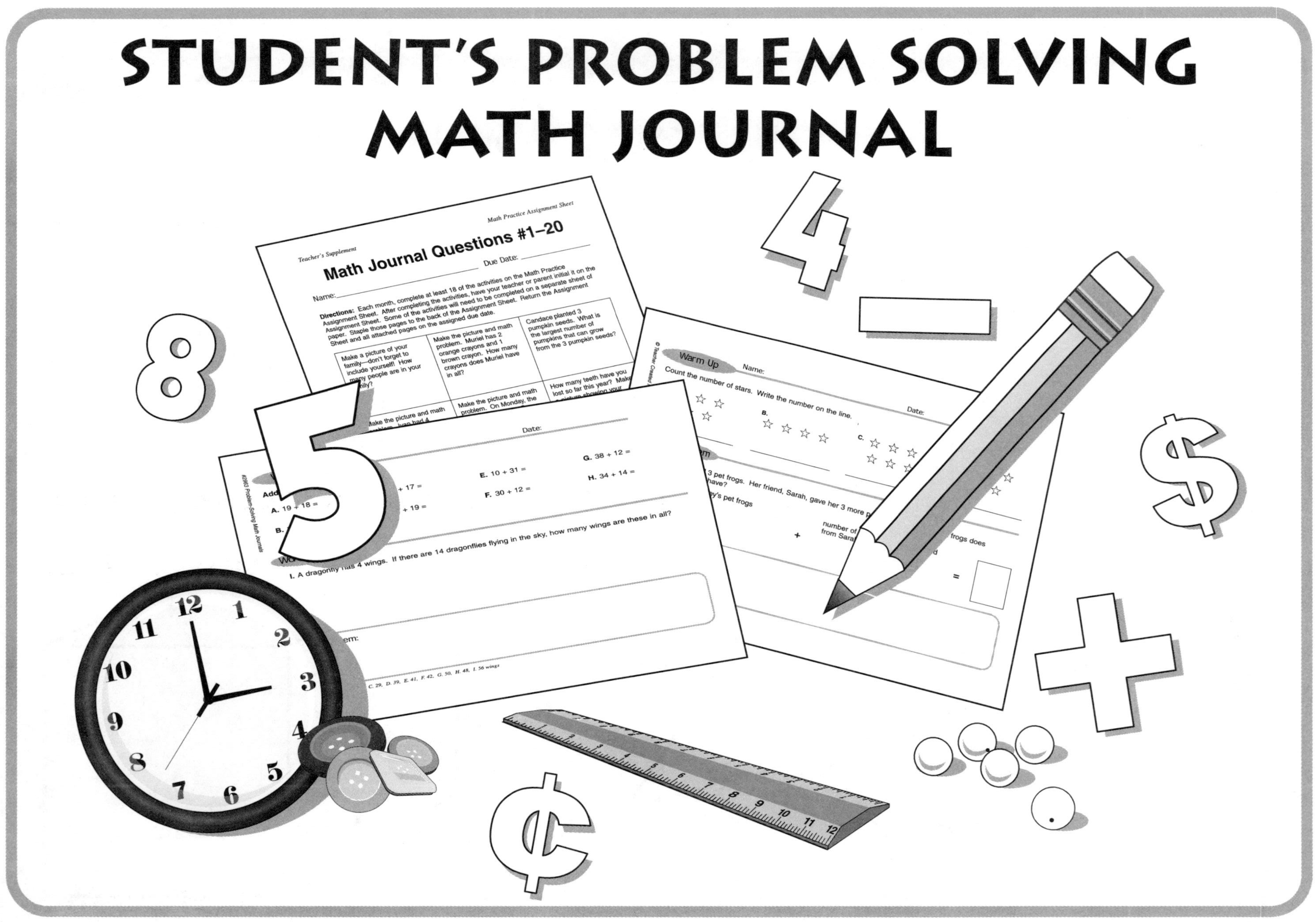
STUDENT'S PROBLEM SOLVING
MATH JOURNAL

Name: Date:

1

Warm Up

Count the number of balloons. Circle the number.

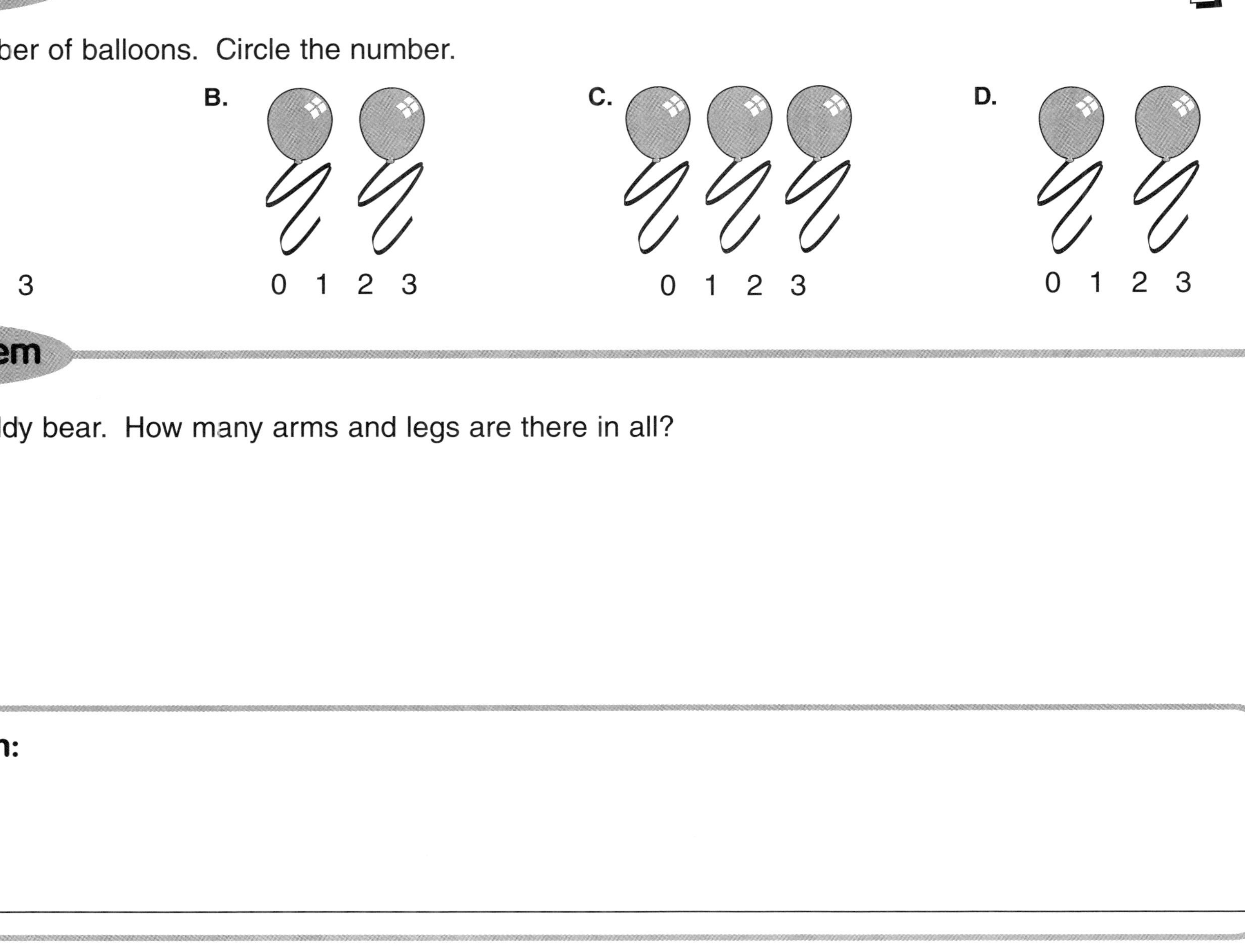

A. 0 1 2 3

B. 0 1 2 3

C. 0 1 2 3

D. 0 1 2 3

Word Problem

E. Draw 1 teddy bear. How many arms and legs are there in all?

Math Problem:

Sentence: ____________________

Answers: A. 1, B. 2, C. 3, D. 2, E. 2 arms and 2 legs, 4 in all

Warm Up

Name: ____________________ Date: ____________________

2

Count the number of crayons. Circle the number.

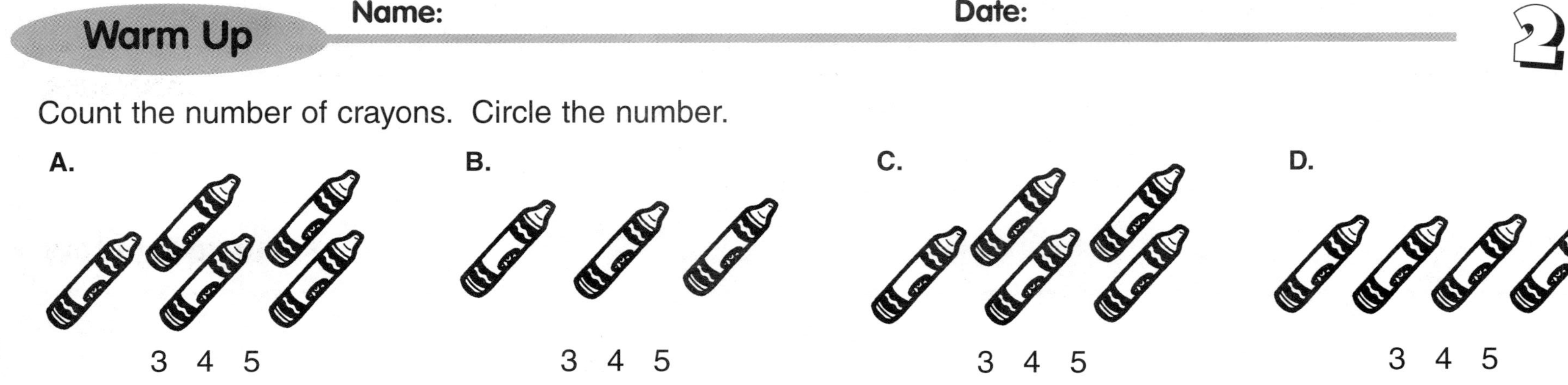

A. 3 4 5

B. 3 4 5

C. 3 4 5

D. 3 4 5

Word Problem

E. A bear saw 2 beehives on Sunday. The bear saw 1 beehive on Monday. How many beehives did the bear see in all?

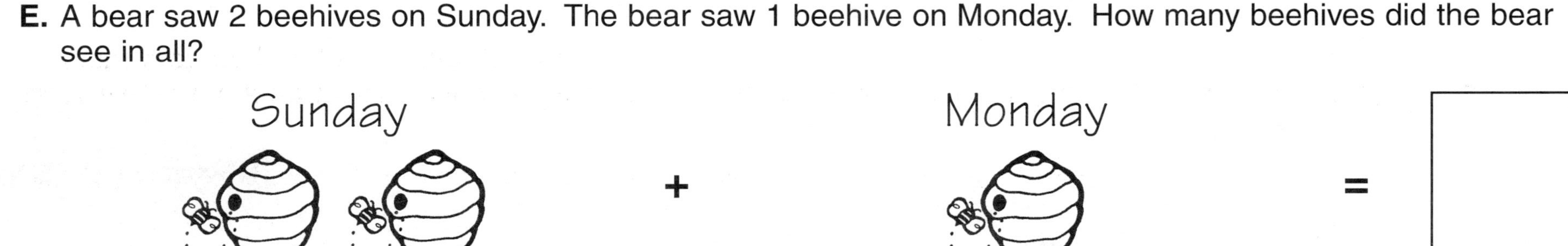

Math Problem:

Sentence: ____________________

Answers: A. 5, B. 3, C. 5, D. 4, E. 3 beehives

3

Warm Up

Name: ____________________ Date: ____________________

Count the number of triangles. Circle the number.

A.

6 7 8

B.

6 7 8

C.

6 7 8

D.

6 7 8

Word Problem

E. Mary picked 2 large pumpkins and 1 small pumpkin. How many pumpkins did Mary pick in all? (Draw a picture of the pumpkins Mary picked.)

Math Problem:

Sentence: ____________________

Answers: A. 6, B. 7, C. 6, D. 8, E. 3 pumpkins

Name: Date:

4

Warm Up

Count the number of squares. Circle the number.

A.

8 9 10

B.

8 9 10

C.

8 9 10

D.

8 9 10

Word Problem

E. The tooth fairy has 6 teeth in her purse. She collected 1 tooth from Anna. She collected 2 teeth from Alina. The rest of the teeth she collected from Jacob. How many teeth did the Tooth Fairy collect from Jacob? (Cut out the correct number of counters from page 251 and glue them below each name.)

Anna Alina Jacob

Math Problem:

Sentence: ____________________

Answers: A. 10, B. 9, C. 8, D. 10, E. 3 teeth

5

Warm Up **Name:** ______ **Date:** ______

Count the letters. Write the number on the line.

A. M m M m ______

B. Q q Q q q ______

C. G ______

D. K k K k K k ______

Word Problem

E. Write the letters of the alphabet. How many letters did you count?

Math Problem:

Sentence: ____________________

Answers: A. 4, B. 5, C. 1, D. 6, E. 26 letters

6

Warm Up

Name: **Date:**

Count the number of legs. Write the number on the line.

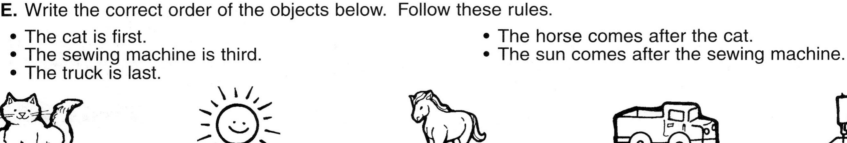

A. ______ **B.** ______ **C.** ______ **D.** ______

Word Problem

E. Write the correct order of the objects below. Follow these rules.

- The cat is first.
- The sewing machine is third.
- The truck is last.
- The horse comes after the cat.
- The sun comes after the sewing machine.

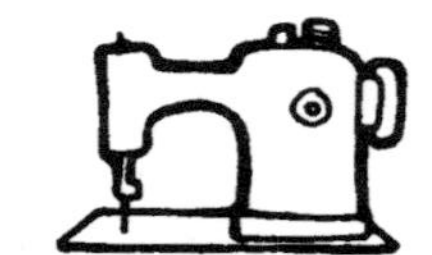

Math Problem:

Sentence: ______

Answers: A. 4, B. 2, C. 0, D. 4, E. cat, horse, sewing machine, sun, and truck

7

Warm Up

Name: **Date:**

Count the number of bees. Write the number on the line.

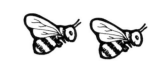

A. **B.** **C.** **D.**

______ ______ ______ ______

Word Problem

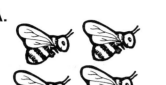

E. How many bunnies are in the circle? ______

F. How many bunnies are not in the circle? ______

G. How many bunnies are there in all? ______

Math Problem:

Sentence: ______________________________

Answers: A. 4, B. 3, C. 0, D. 2, E. 5, F. 3, G. 8

Name: **Date:** 8

Warm Up

Count the number of flowers. Write the number on the line.

A. ________

B. ________

C. ________

D. ________

Word Problem

E. Veronica had 5 pet sheep. She gave 2 of her sheep to Avery. How many sheep does Veronica have left? Circle the correct number of sheep that Veronica has left.

Math Problem:

Sentence: ____________________

Answers: A. 3, B. 2, C. 5, D. 3, E. 3

Warm Up

Name: ______________________ Date: ______________________

9

Draw the number of footballs.

2 **4** **1** **3**

Word Problem

A. How many footballs are in the circle? ______________

B. How many footballs are not in the circle? ______________

C. How many baseballs are in the circle? ______________

D. How many baseballs are not in the circle? ______________

Math Problem:

Sentence: ____

Answers: A. 3, B. 5, C. 0, D. 3

Name: ______ **Date:** ______ **10**

Warm Up

Count the number of school bells. Write the number on the line.

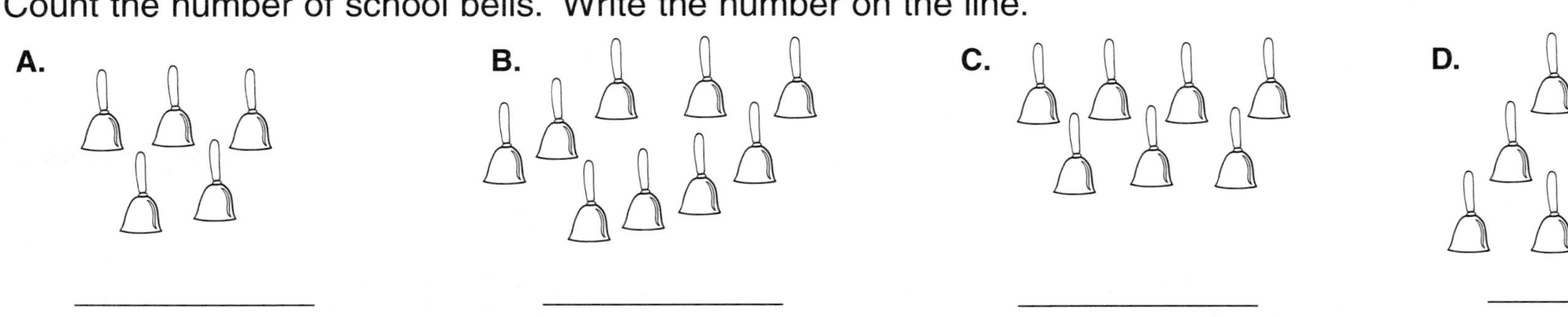

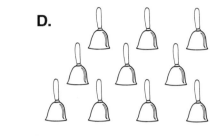

A. ______ **B.** ______ **C.** ______ **D.** ______

Word Problem

E. Veronica read 5 books in March and 4 books in April. Color the correct number of books Veronica read in March blue and then color the total number of books she read in April orange.

Math Problem:

Sentence: ______

Answers: A. 5, B. 9, C. 7, D. 10, E. blue—5 books, orange—4 books

Warm Up

Name: **Date:** 11

Count the number of stars. Write the number on the line.

A. ☆ ☆ ☆ ☆ ☆ ________

B. ☆ ☆ ☆ ☆ ________

C. ☆ ☆ ☆ ☆ ☆ ☆ ________

D. ☆ ☆ ☆ ☆ ☆ ________

Word Problem

E. Courtney had 3 pet frogs. Her friend Sarah gave her 3 more pet frogs. How many pet frogs does Courtney have now?

Courtney's pet frogs

+

number of frogs Courtney received from Sarah

=

Math Problem:

Sentence: ________________________________

Answers: A. 5, B. 4, C. 6, D. 5, E. 6

12

Warm Up

Name: ______________________ Date: ______________________

Trace each number.

0 1 2 3 4 5 6 7 8 9 10

Word Problem

Louisa and Joanna had a tea party. They each brought 4 dolls to the party. How many dolls were there in all? Draw the number of dolls that Louisa and Joanna brought to the tea party. Glue the correct number of counters from page 251 to show the correct number of dolls.

Math Problem:

Sentence: ______________________

Answer: There were 8 dolls at the tea party.

Name: ______________________ **Date:** ______________________

13

Warm Up

Write the numbers 0 to 10.

Word Problem

Allie saw 5 pumpkins on the fence. Cori saw 4 pumpkins on the ground. How many pumpkins did Allie and Cori see in all?

Math Problem:

Sentence: ______________________

Answer: Allie and Cori saw 9 pumpkins.

14

Warm Up

Name: Date:

Read each number word. Trace each number word.

zero one two three

Word Problem

One week has 7 days. It was sunny 4 days last week. The rest of the week it rained. How many days did it rain last week? (*Draw suns on 4 of the days. Draw rain on the remaining days.*)

Sunday	Monday	Tuesday	Wednesday	Thursday	Friday	Saturday

Math Problem:

Sentence: ______________________

Answer: It rained 3 days.

15

Warm Up

Name: Date:

Read each number word. Trace each number word.

four five six seven

Word Problem

The rabbit lives in the first house. The squirrel lives in the last house. The robin lives in the third house. The duck lives in the second house. Glue the correct picture of each animal from page 251 to the correct house.

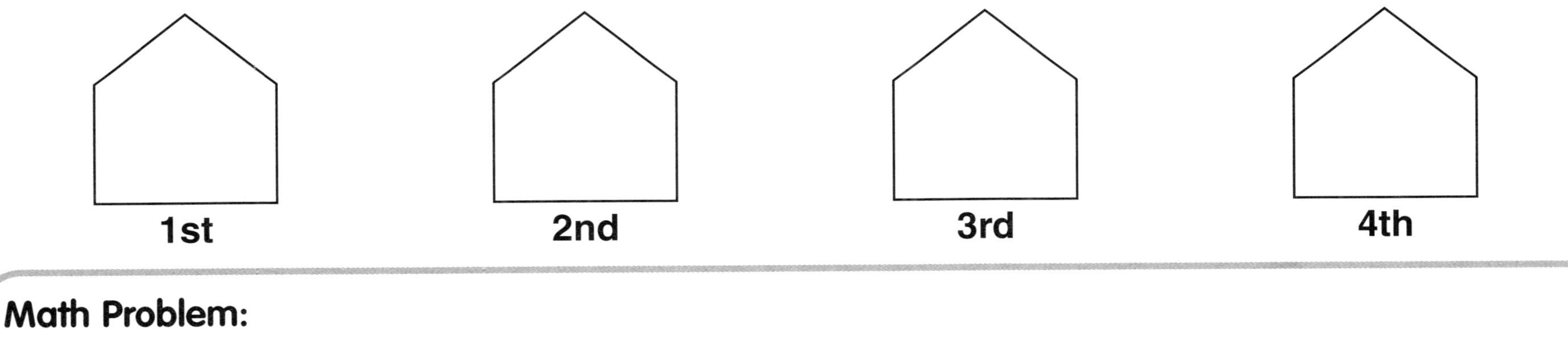

Math Problem:

Sentence: ____________________

Answer: Order of where animals live—rabbit, duck, robin, and squirrel.

16

Warm Up

Name: ____________ **Date:** ____________

Read each number word. Trace each number word.

eight   nine   ten

Word Problem

A tiger has 4 legs. If there are 3 tigers, how many legs are there in all? ____________

 ____________

 ____________

____

Math Problem:

Sentence: ____________

Answer: There are a total of 12 legs in all.

17

Warm Up

Name: ____________________ Date: ____________________

Follow the directions.

Color 1 shoe.

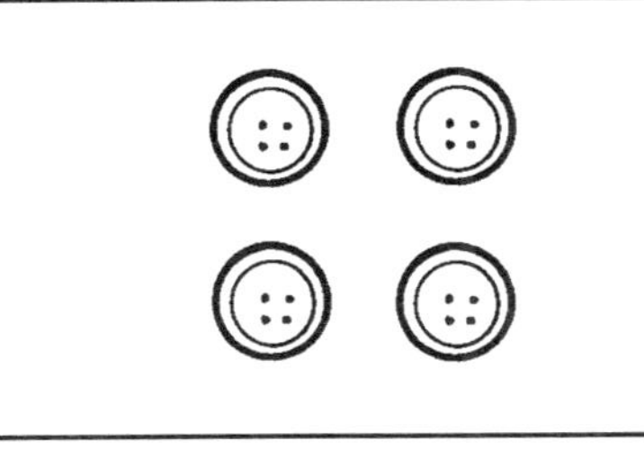

Color 2 buttons.

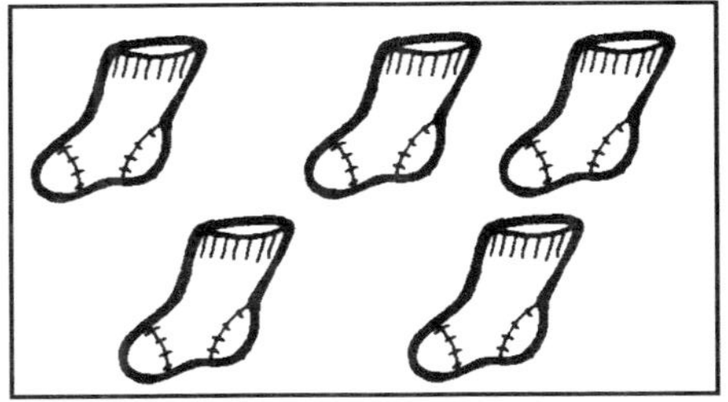

Color 3 socks.

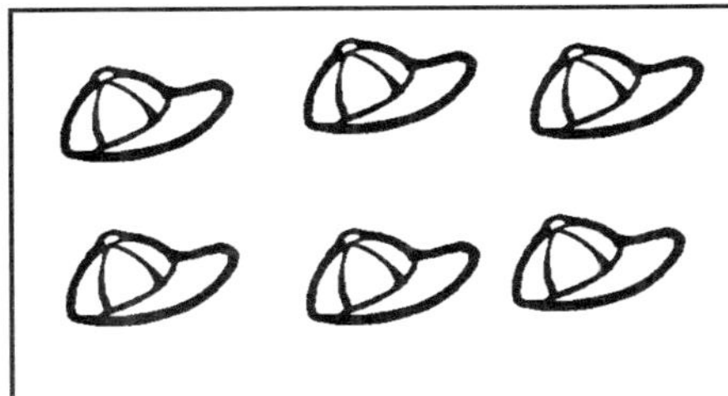

Color 4 hats.

Word Problem

Color 1 pair of shorts red and 1 pair of shorts yellow. Color 1 shirt green and 1 shirt purple. How many different outfits can you make with the shorts and shirts you colored?

shorts + **shirts**

Math Problem:

Sentence: ____________________

Answer: Different combinations include—red shorts + green shirt; red shorts + purple shirt; yellow shorts + green shirt; yellow shorts + purple shirt

Warm Up

Name: ______ **Date:** ______

18

Draw the number of circles.

5	6	7	8

Word Problem

A. How many mice are in the square? ______

B. How many mice are not in the square? ______

C. How many mice are there in all? ______

Math Problem:

Sentence: ______

Answers: A. 6, B. 4, C. 10

19

Warm Up

Name: ______ Date: ______

Draw one of each shape.

circle	square	triangle	oval

Word Problem

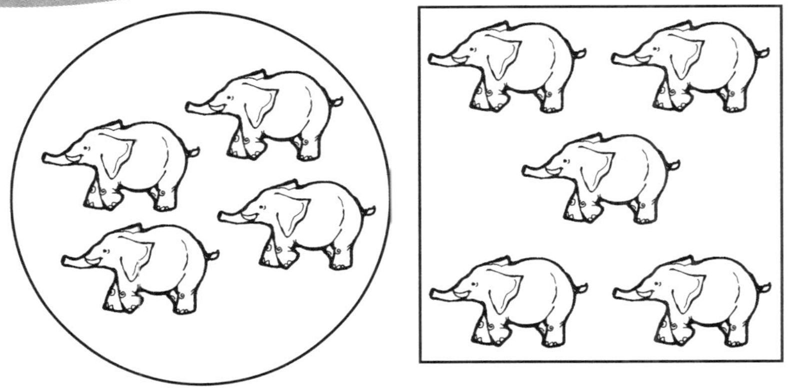

A. How many elephants are there in the circle? ______

B. How many elephants are there in the square? ______

C. How many elephants are there in all? ______

Math Problem:

Sentence: ______

Answers: A. 4, B. 5, C. 9

20

Warm Up

Name: ______________________ Date: ______________________

Complete each pattern.

A. G H G H ____ ____ **B.** 1 2 1 2 1 ____ ____ **C.** H h H h H ____ ____ **D.** P 3 P 3 P ____ ____

Word Problem

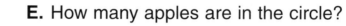

E. How many apples are in the circle? ____________

F. How many pineapples are in the square? ____________

G. How many pineapples and apples are not in the square and the circle? ____________

Math Problem:

Sentence: ____

Answers: A. G, H B. 2, 1 C. h, H D. 3, P E. 2, F. 3, G. 4

21

Warm Up **Name:** ______ **Date:** ______

Read each number word. Write the number on the line.

zero	one	two	three	four
____	____	____	____	____

Word Problem

Which of the three items is the longest?

Math Problem:

Sentence: ______________________

Answer: The pencil is longest

Name: Date:

22

Warm Up

Read each number word. Write the number on the line.

four ________ five ________ six ________ seven ________ eight ________

Word Problem

Maria invited people to her birthday party. She invited 6 friends on Wednesday and 10 friends on Thursday. How many people did Maria invite to her birthday party in all? Color in the correct number of squares for each day.

Wednesday	1	2	3	4	5	6	7	8	9	10
Thursday	1	2	3	4	5	6	7	8	9	10

Math Problem:

Sentence: ________________________________

Answer: 4; 5; 6; 7; 8; Maria invited 16 people to her party.

23

Warm Up

Name: ______________________ Date: ______________________

Read each number word. Write the number on the line.

six	seven	eight	nine	ten
________	________	________	________	________

Word Problem

Write your favorite number word. ____________

One monkey has 2 eyes. If there are 4 monkeys, how many eyes would there be in all? ____________ eyes

____________ eyes ____________ eyes ____________ eyes ____________ eyes

Math Problem:

Sentence: ____

Answer: There are 8 eyes.

24

Warm Up

Name: ____________________ Date: ____________________

Read each number. Write the number word on the line.

0 ____________ 2 ____________ 4 ____________

1 ____________ 3 ____________ 5 ____________

Word Problem

When Luis looked out in his backyard, he found all of these animal footprints. If each animal has 4 legs and each leg makes a print, how many animals were in Luis' backyard?

Math Problem:

Sentence: ____

Answer: There were 6 animals in Luis' backyard.

25

Warm Up

Name: ____________ Date: ____________

Read each number. Write the number word on the line.

4 ____________ 6 ____________ 8 ____________

5 ____________ 7 ____________ 9 ____________

Word Problem

Count how many leaves are in the picture below. ____________ Are they different or similar? ____________

Math Problem:

Sentence: ____

Answer: There are 10 leaves that are different.

Warm Up

Name: Date:

26

Read each number. Write the number word on the line.

8 ______	10 ______	1 ______
9 ______	0 ______	2 ______

Word Problem

Carlo bought some baseball caps. He bought 2 caps every week. If there are 4 weeks in one month, how many baseball caps did Carlo buy in one month? (Glue the total number of baseball caps that Carlo bought from the counters provided on page 251.)

Math Problem:

Sentence: ______

Answer: Carlo bought 8 caps in all.

27

Warm Up

Name: Date:

Trace each number.

0 1 2 3 4 5 6 7 8 9 10

Word Problem

There were 8 bear cubs in the forest, and 3 of the cubs were asleep in a cave. How many cubs were not sleeping in the cave? Cut out the correct number of counters from page 251 and glue them in the space below to show how many cubs were not sleeping in the cave.

Math Problem:

Sentence: ____________________

Answer: There were 5 cubs not in the cave.

28

Warm Up

Name: Date:

Write the missing numbers.

10, 9, ___, 7, 6, ___, 4, 3, ___, 1, 0

Word Problem

There are 2 bears. How many different ways could you divide 3 jars of honey between the two bears?

bear #1		bear #2		
0 jars	+	3 jars	=	3 jars
______	+	______	=	3 jars
______	+	______	=	3 jars
______	+	______	=	3 jars

Math Problem:

Sentence: ____

Answer: Possible solutions include—1 jar + 2 jars, 2 jars + 1 jar, 3 jars + 0 jars

Name: Date: **29**

Warm Up

Write the numbers 0 to 10.

Word Problem

Using red and green crayons, color the apples to make an "AB" pattern. Use the letters "A" and "B" to label the pattern. Write a math problem that tells about the number of red apples and the number of green apples in the pattern. Write a sentence describing the pattern.

Math Problem:

Sentence: ____________________

Warm Up

Name: **Date:** **30**

Look at the boxes. Answer the questions.

☆ ☆ ☆ How are the 3 shapes alike?	□ ○ △ How are the 3 shapes different?

Word Problem

Look at the 3 boxes. How are they alike? How are they different?

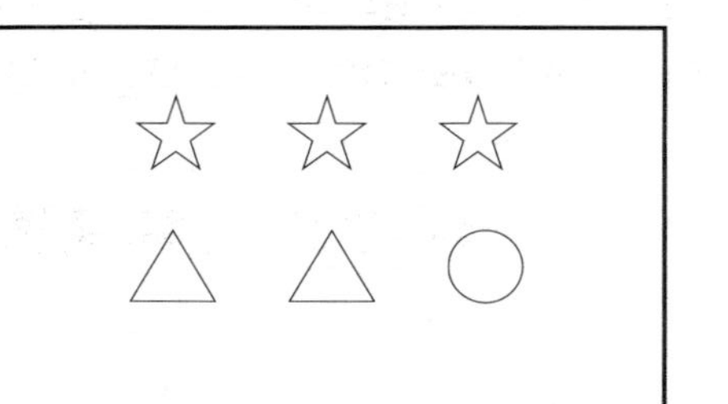

Math Problem:

Sentence: ____________________

Warm Up

Name: ________________ **Date:** ____________

Write the missing numbers.

Word Problem

The Thompson family went fishing on Friday. They caught 12 big fish. They ate 6 of the fish for dinner. How many fish are left?

Math Problem:

Sentence: ____________________________

Answer: There are 6 fish left.

Warm Up

Name: **Date:**

32

Write the number that comes in the middle.

A. 13, _____, 15 **B.** 0, _____, 2 **C.** 14, _____, 16 **D.** 18, _____, 20

Word Problem

E. Patti collected seashells at the beach. If she found 2 seashells every day, how many seashells did Patti have at the end of the week?

Sunday	Monday	Tuesday	Wednesday	Thursday	Friday	Saturday

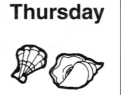

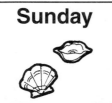

Math Problem:

Sentence: ________________________________

Answer: A. 14, B. 1, C. 15, D. 19, E. 14 seashells

Name: ________ **Date:** ________

33

Warm Up

Write the number that comes before.

A. _____, 15, 16   **B.** _____, 5, 6   **C.** _____, 8, 9   **D.** _____, 13, 14

Word Problem

E. Jamal needs 35¢ to buy an apple. What is the correct combination of coins that can be used to make 35¢? (Glue the correct number of money counters provided on page 251.)

Math Problem:

Sentence: ____________________

Answers: A. 14, B. 4, C. 7, D. 12, E. 3 dimes + 1 nickel

Warm Up

Name: Date:

34

Fill in the missing numbers.

A. ____, 13, 14, ____ **B.** 16, 17, ____, ____ **C.** 3, ____, ____,6 **D.** 16, ____, 18, ____

Word Problem

E. Find your state on a map of the United States. How many other states touch your state?

Draw the shape of your state below.

Math Problem:

Sentence: ____________________

Answers: A. 12, 15, B. 18, 19, C. 4, 5, D. 17, 19, E. Answer will vary.

Warm Up

Name: ____________ **Date:** ____________

35

Fill in the missing numbers.

A. 14, ____, 16, ____ **B.** 17, ____, ____, 20 **C.** 15, ____, 17, ____ **D.** 6, ____, ____, 9

Word Problem

E. Marcus has 3 pet birds and 4 dogs. Sally has 4 pet birds and 3 dogs. How many pets do they have in all? (Using the counters provided on page 251, glue the correct number of animals that each person has.)

Marcus + Sally = ☐

Math Problem:

Sentence: ____________

Answers: A. 15, 17 B. 18, 19, C. 16, 18, D. 7, 8, E. 14 pets

Name: ______________________ Date: ______________________

36

Warm Up

Subtract.

A. 13 – 3 = ________

B. 14 – 6 = ________

C. 15 – 4 = ________

D. 11 – 1 = ________

Word Problem

E. Marcie just loves fish! As a matter of fact, she has quite a collection of fish in her fish tank. Look at the fish in Marcie's fish tank. Write a math problem and a sentence about the fish.

Math Problem:

Sentence: ____

Answers: A. 10, B. 8, C. 11, D. 10, E. Answers will vary.

37

Warm Up

Name: Date:

Write the number that comes next.

A. 0, 1, _____ **B.** 3, 4, _____ **C.** 6, 7, _____ **D.** 2, 3, _____

Word Problem

E. Harvey can put 6 seahorses in 1 aquarium. Harvey has 3 aquariums. How many seahorses does Harvey have? (Using the counters provided on page 252, glue the correct number of seahorses in each aquarium.)

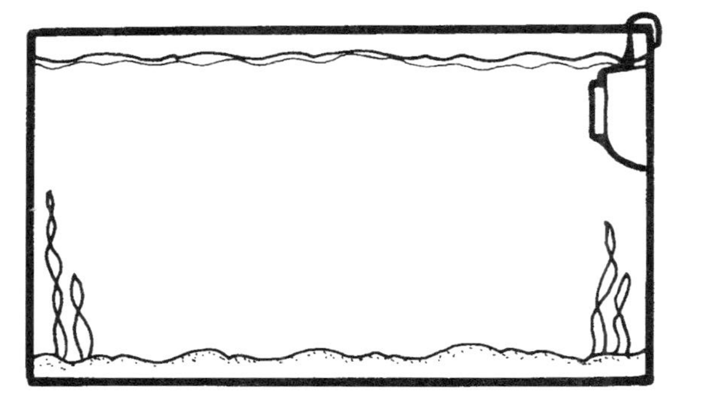

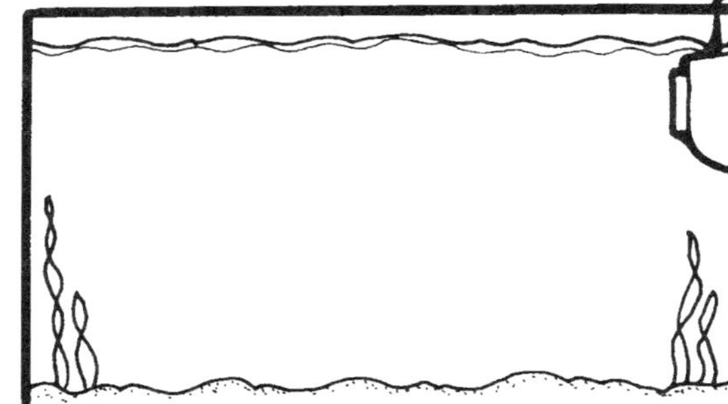

Math Problem:

Sentence: ______________________________

Answers: A. 2, B. 5, C. 8, D. 4, E. 18 seahorses

Warm Up

Name: ______________________ Date: ______________________ **38**

Fact Families: Use 2, 3, and 5 to make 2 addition problems and 2 subtraction problems.

A. ☐ + 3 = 5

B. ☐ + 2 = 5

C. ☐ – 3 = 2

D. 5 – ☐ = 3

Word Problem

E. Zoey has 3 wagons and 27 apples. How many apples does Zoey need to put in each wagon so that all of the wagons have the same number of apples? Draw the correct number of apples that Zoey puts in each wagon.

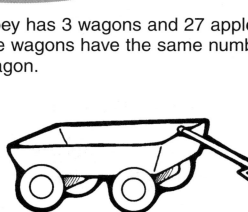

Math Problem:

Sentence: ______________________

Answers: A. 2, B. 3, C. 5, D. 2, E. 9 apples

39

Warm Up

Name: ____________ Date: ____________

Look at each number. Make the correct number of tally marks. The first one has been done for you.

////			
4	11	5	3

Word Problem

Solve the math pattern and write a math problem and sentence about the pattern below.

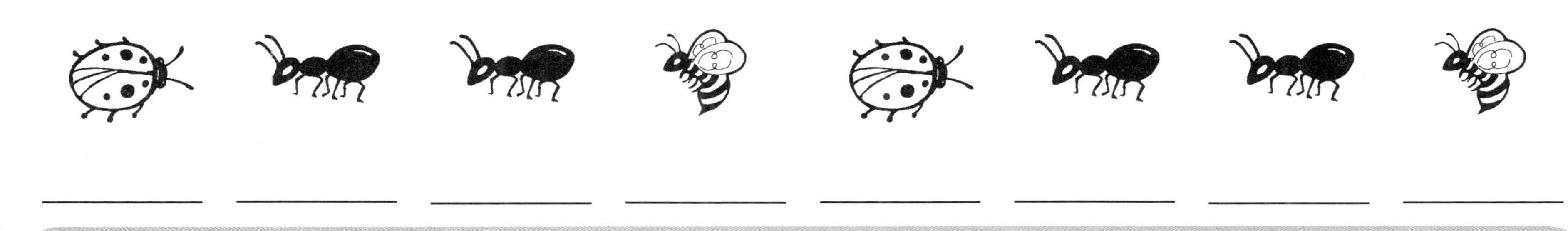

Math Problem:

Sentence: ____________________

Answers: A, B, B, C pattern

40

Warm Up

Name: Date:

Add.

A. 11 + 3 = ________ **C.** 14 + 1 = ________

B. 8 + 2 = ________ **D.** 3 + 5 = ________

Word Problem

E. At the corner grocery store, there is a special on oranges—4 for $1.00. If Wendy bought $3.00 of oranges, how many oranges did she buy?

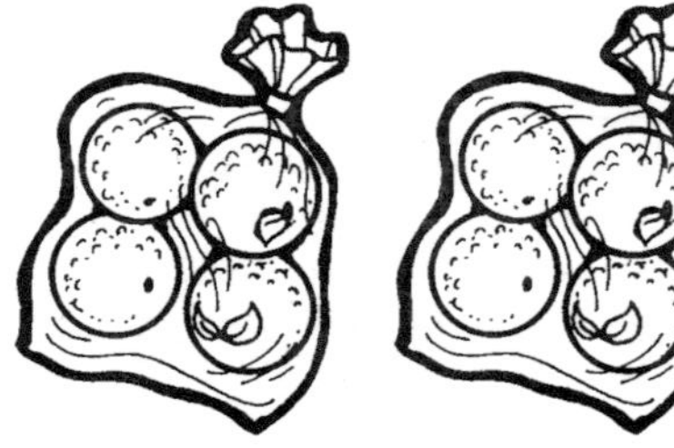

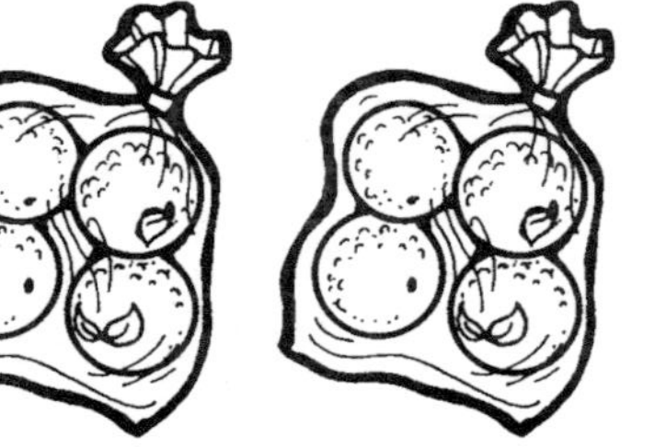

__4__ oranges for $1 ________ oranges for $2 ________ oranges for $3

Math Problem:

Sentence: ____

Answers: A. 14, B. 10, C. 15, D. 8, E. 12 oranges

41

Warm Up

Name: ______________________ Date: ______________________

Read each riddle. Write the answer.

A. I am the number after 0. ☐

B. I am the number before 4. ☐

C. I am the number before 1. ☐

D. I am the number after 1. ☐

Word Problem

E. Clue #1: The Morris family has more than 3 cats, but fewer than 7 cats.
Clue #2: Together, all of the cats have 5 tails.
How many cats does the Morris family have? (Glue the correct number of cats from the counters provided on page 252.)

Math Problem:

Sentence: ______________________

Answers: A. 1, B. 3, C. 0, D. 2, E. 5 cats

Warm Up

Name: Date:

42

Solve each riddle. Write the answer.

A. I am the number before 10. ☐

B. I am the number after 7. ☐

C. I am the number before 5. ☐

D. I am the number after 6. ☐

Word Problem

E. Jamie's horse ate 4 carrots on Sunday. Jamie's horse ate 3 carrots on Monday. How many carrots did Jamie's horse eat in all? Glue the correct number of carrots provided as counters on page 252.

Sunday:

Monday:

Math Problem:

Sentence: ______________________________

Answers: A. 9, B. 8, C. 4, D. 7, E. 7 carrots

Warm Up

Name: Date:

43

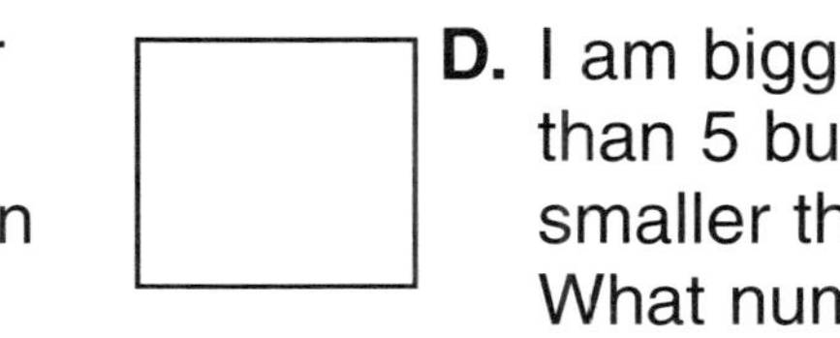

Read each clue. Write the number.

A. I am bigger than 2 but smaller than 4. What number am I?

B. I am bigger than 0 but smaller than 2. What number am I?

C. I am bigger than 8 but smaller than 10. What number am I?

D. I am bigger than 5 but smaller than 7. What number am I?

Word Problem

E. Samuel had 10 toys. He gave half of the toys to Margaret. How many toys does Samuel have left?

Math Problem:

Sentence: ______________________________

Answers: A. 3, B. 1, C. 9, D. 6, E. 5 toys

Warm Up **Name:** _______________ **Date:** _______________ **44**

Follow the directions.

Write a number larger than 10. __________

Write a number smaller than 10. __________

Write a number between 0 and 10. __________

Write a number between 11 and 20. __________

Word Problem

Write the numbers 10 to 20. How many 1's did you write?

Math Problem:

Sentence: ______________________________

Answer: 11 ones

Warm Up

Name: **Date:**

How many legs can be seen on each creature? Count the legs. Write the number of legs on the lines.

A.  ____________

B.  ____________

C. ____________

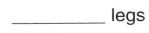

D. ____________

Word Problem

E. A spider has 4 legs on the left side of its body. A spider has the same number of legs on the right side of its body. How many legs do 4 spiders have in all?

 ________ legs

 ________ legs

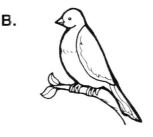

 ________ legs

 ________ legs

Math Problem:

Sentence: ____

Answers: A. 6, B. 2, C. 4, D. 4, E. 32 legs

Warm Up

Name: __________ Date: __________

46

Add or subtract.

A. 15 – 3 = ________

B. 16 + 0 = ________

C. 14 – 2 = ________

D. 13 + 3 = ________

Word Problem

E. A grasshopper is 2" (5 cm) long. An earthworm is 4" (10 cm) longer than the grasshopper. How long is the earthworm?

Math Problem:

Sentence: ____________________

Answers: A. 12, B. 16, C. 12, D. 16, E. 6 inches (15 cm)

Name: Date: **47**

Warm Up

Subtract.

A. 18 – 3 = ________

B. 15 – 5 = ________

C. 18 – 7 = ________

D. 17 – 6 = ________

Word Problem

Use circles, squares, triangles, ovals, and rectangles to design a face for a pumpkin. Write a math problem that describes the number of different shapes used in the design. Write a sentence describing the design.

Math Problem:

Sentence: ____

Answers: A. 15, B. 10, C. 11, D. 11

Warm Up

Name: Date:

A. Write an addition sentence for the hearts and stars.

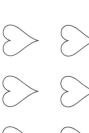

________ + ________ = ________
hearts stars shapes

♡ ♡ ♡ ☆ ☆ ☆
♡ ♡ ♡ + ☆ ☆ ☆

Word Problem

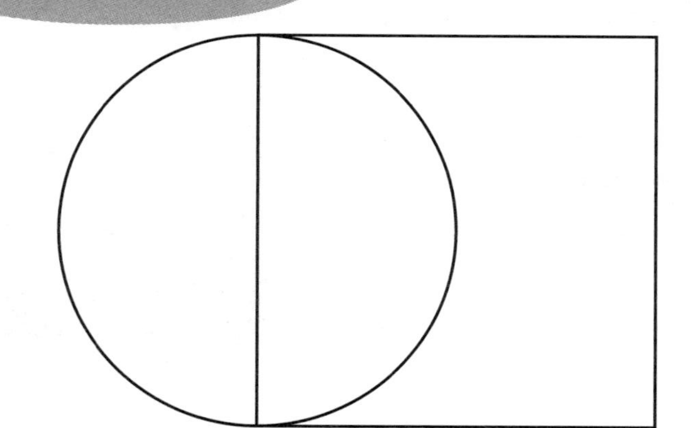

Draw 2 hearts in the circle but not in the square. Draw 3 stars in the square but not in the circle. Draw 1 heart in both the circle and the square. Draw 1 star in both the circle and the square.

B. How many shapes did you use?

C. How many shapes are in both the circle and the square?

Math Problem:

Sentence: ____________________

Answers: A. 6 hearts + 6 stars = 12 shapes, B. 7, C. 2

Warm Up

Name: **Date:**

1. Circle the 3rd cup. **2.** Put an **X** on the 8th cup. **3.** Draw a line under the 1st cup.

Word Problem

A. How many days are there between the 5th of August and the 12th of August? (Don't count August 5th or August 12th—only the days in between them.) ________

B. How many Mondays are in the month of August?________

C. What is the last day of the week in the month of August?________

August						
S	M	T	W	Th	F	S
		1	2	3	4	5
6	7	8	9	10	11	12
13	14	15	16	17	18	19
20	21	22	23	24	25	26
27	28	29	30	31		

Math Problem:

Sentence: ________________________________

Answers: A. 6 days, B. 4, C. Thursday

Warm Up

Name: ______________________ Date: ______________________

Cross out 4. How many are left? Write the answer on the line.

A.  ________

B.  ________

C.  ________

D. ________

Word Problem

E. Sarah saw 12 seahorses, but 4 of the seahorses swam away. How many seahorses were left? Cross out the seahorses that swam away.

Math Problem:

Sentence: ______________________

Answers: A. 4, B. 2, C. 3, D. 0, E. 8 sea horses

51

Warm Up **Name:** **Date:**

Circle the longest line.

A.

B.

Circle the shortest line.

C.

D.

Word Problem

E. A fir tree is 7 feet tall (2 m). An oak tree is 2 feet taller (3 m).

Which is the taller tree? ________

Which is the shorter tree? ________

Math Problem:

Sentence: ________________________________

Answers: A. third line, B. second line, C. first line, D. second line, E. The oak tree is the taller tree, and the fir tree is the shorter tree.

Warm Up

Name: ______________________ Date: ______________

Solve each riddle.

A. I am one foot long. I measure things. What am I?

____

B. I say “tick-tock.” I hang on the wall. What am I?

____

C. I can be used to buy things. I am worth 1¢ What am I?

____

D. I am on each hand. There are 10 of me in all. What am I?

____

Word Problem

E. A watermelon was cut into 12 slices. Seth ate 4 slices of watermelon. Color the correct number of watermelon slices that Seth didn’t eat.

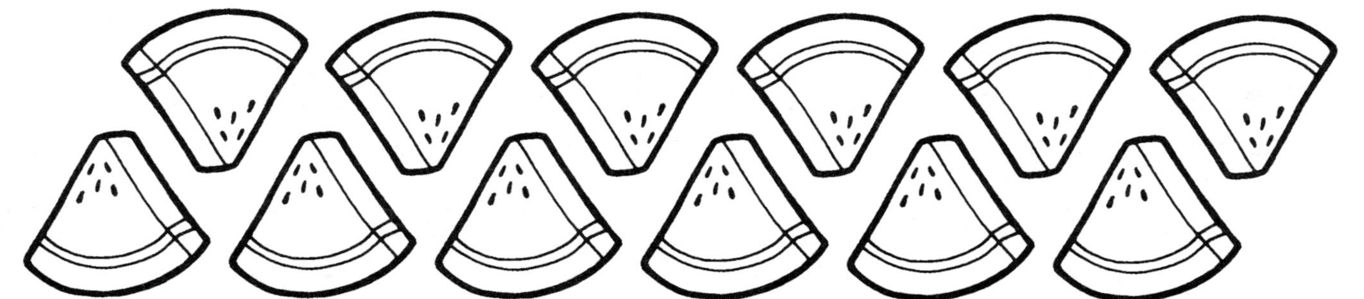

Math Problem:

Sentence: ______________________

Answers: A. ruler, B. clock, C. penny, D. finger, E. 8 watermelon slices

53

Warm Up

Name: Date:

Add or subtract.

A. $\begin{array}{r} 18 \\ -\ 7 \\ \hline \end{array}$

B. $\begin{array}{r} 5 \\ +\ 13 \\ \hline \end{array}$

C. $\begin{array}{r} 8 \\ +\ 10 \\ \hline \end{array}$

D. $\begin{array}{r} 17 \\ -\ 4 \\ \hline \end{array}$

Word Problem

E. Ben collects leaves. He has 8 large leaves. He has 7 small leaves. How many leaves does Ben have in his collection? Glue the correct number of leaves, using the counters provided on page 252.

Math Problem:

Sentence: ______________________________

Answers: A. 11, B. 18, C. 18, D. 13, E. 15 leaves

54

Warm Up

Name: ______ Date: ______

Add.

A. 6 + 1 = ______

B. 11 + 8 = ______

C. 15 + 0 = ______

D. 0 + 8 = ______

E. 12 + 5 = ______

F. 15 + 12 = ______

G. 12 + 2 + 2 + 1 = ______

H. 0 + 11 + 3 + 5 = ______

Word Problem

I. Billy saw 5 bats flying in a cave. Laura saw 8 bats hanging upside down in the cave. How many bats did Billy and Laura see in all? Using the counters provided on page 252, glue the correct number of bats.

Math Problem:

Sentence: ______

Answers: A. 7, B. 19, C. 15, D. 8, E. 17, F. 27, G. 17, H. 19, I. 13 bats

55

Warm Up

Name: ______________________ **Date:** ______________________

Look at each shape. Count the corners. Write the number on the line.

A. 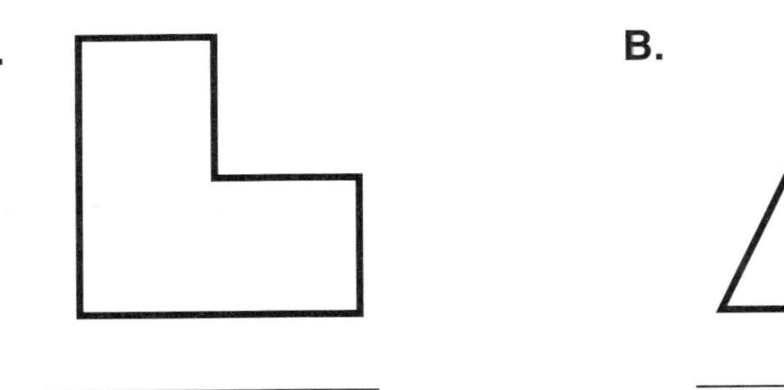 ________

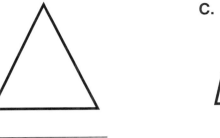

B. ________

C. 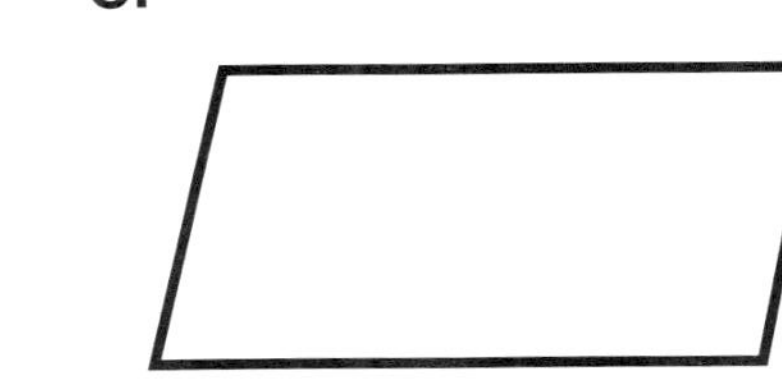  ________

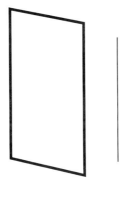

D. ________

Word Problem

Design a flag using circles, squares, stars, hearts, ovals, rectangles, and any other simple shape you like. Write a math problem and a sentence describing the flag you made.

Math Problem:

Sentence: ____

Answers: A. 6, B. 3, C. 4, D. 0

Warm Up

Name: Date: **56**

Look at the numbers. Answer the questions.

A. How are the numbers alike?	B. How are the numbers different?
7 7 7	7 2 5

Word Problem

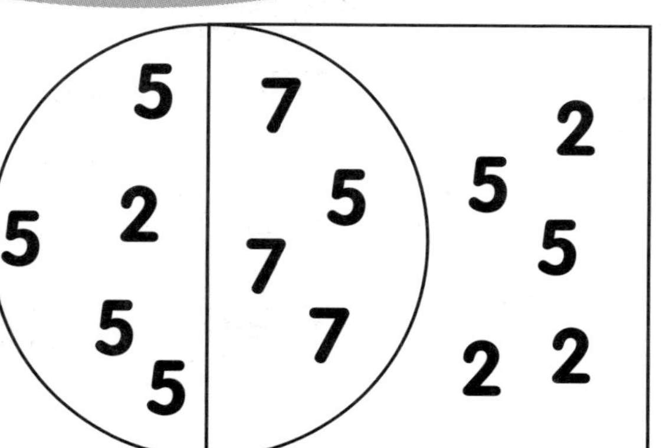

C. How many 7's are in the circle? ________

D. How many 5's are in the circle but not the square? ________

E. How many 2's are in the square? ________

F. How many 7's are in the square but not the circle? ________

G. Which number is in the circle, the square, and in both the circle and the square? ________

Math Problem:

Sentence: ________________

Answers: A. all the numbers are 7, B. all the digits are different numbers, C. 3, D. 4, E. 3, F. 0, G. 5

57

Warm Up

Name: **Date:**

Write the number that is 4 more.

A. 1 + ________ = ________

C. 3 + ________ = ________

B. 2 + ________ = ________

D. 4 + ________ = ________

Word Problem

E. A cougar has 4 legs. How many legs would 4 cougars have? Count the total number of legs in the picture below.

________ ________ ________ ________

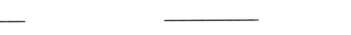

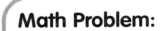

Math Problem:

Sentence: ______________________________

Answers: A. 5, B. 6, C. 7, D. 8, E. 16 legs

Warm Up

Name: **Date:**

58

Fact Families: Use 3, 4, and 7 to make 2 addition and 2 subtraction problems.

A. 3 + ________ = ________

B. 4 + ________ = ________

C. ________ − 3 = ________

D. 7 − ________ = ________

Word Problem

E. There are many animals at the zoo. There are 5 hippos, 7 seals, and 6 walruses swimming in a big pool. How many animals are there in the big pool? Glue the correct number of counters provided on page 253.

Math Problem:

Sentence: ____________________

Answers: A. 3 + 4 = 7, B. 4 + 3 = 7, C. 7 − 3 = 4, D. 7 − 4 = 3, E. 18 animals

Name: **Date:** **59**

Warm Up

Look at each pair of numbers. Circle the number that is smaller.

A. 9 6

C. 8 17

B. 11 5

D. 12 4

Word Problem

E. Molly has 20 oranges. Michael has 8 oranges. How many oranges do Molly and Michael have together? Glue the correct number of counters provided on page 253.

Math Problem:

Sentence: ______________________________

Answers: A. 6, B. 5, C. 8, D. 4, E. 28 oranges

Warm Up

Name: ____________________ **Date:** ____________________

60

Write 4 addition and/or subtraction problems that have 8 as the answer.

_____ ☐ _____ = 8 _____ ☐ _____ = 8

_____ ☐ _____ = 8 _____ ☐ _____ = 8

Word Problem

The Brown family collects feathers. Mr. Brown has 15 feathers. Mrs. Brown has 13 feathers. Lucy Brown has 8 feathers. How many feathers does the Brown family have in all?

Mr. Brown **Mrs. Brown** **Lucy Brown**

Math Problem:

Sentence: ____________________

Answer: The Brown family has 36 feathers.

Name: Date:

61

Warm Up

Add.

A. 22 + 16

B. 15 + 8

C. 5 + 21

D. 16 + 12

Word Problem

E. Gina and Gary went fishing. Gary caught a fish that was 4" (10 cm) long. Gina's fish was 2" (5 cm) longer than Gary's. How long was Gina's fish? (Draw picture of Gina's fish.)

Math Problem:

Sentence: ____________________

Answers: A. 38, B. 23, C. 26, D. 28, E. 6" (15 cm)

Name: Date:

62

Warm Up

Write 4 addition and/or subtraction problems that have 12 as the answer.

______ ☐ ______ = ______   ______ ☐ ______ = ______

______ ☐ ______ = ______   ______ ☐ ______ = ______

Word Problem

Nicole works in a shoe shop. She just unpacked a shipment of shoes. Draw lines to help Nicole match each pair of shoes.

Math Problem:

Sentence: ____________________________

63

Warm Up

Name: ____________________ Date: ____________________

Add or subtract.

A. 16 + 11 = ____

B. 18 − 5 = ____

C. 14 + 4 = ____

D. 28 − 13 = ____

Word Problem

E. The Sampson family went out for a walk in the woods. The Sampsons saw 15 big oak trees and 8 small oak trees. How many oak trees did the Sampson family see in all? Using the counters on page 253, glue the correct number of trees the Sampson family saw in the woods.

Math Problem:

Sentence: ____________________

Answers: A. 27, B. 13, C. 18, D. 15, E. 23 oak trees

64

Warm Up

Name: ____________ Date: ____________

Add.

A. 32 + 11

B. 10 + 16

C. 24 + 25

D. 12 + 14

Word Problem

E. Adam saw 12 squirrels, 4 bears, and 3 deer. How many animals did Adam see in all?

Math Problem:

Sentence: ____________

Answers: A. 43, B. 26, C. 49, D. 26, E. 19 animals

65

Warm Up

Name: Date:

Subtract.

A. 19 – 10 = ________ C. 23 – 10 = ________ E. 19 – 0 = ________ G. 11 – 0 = ________

B. 13 – 3 = ________ D. 17 – 3 = ________ F. 15 – 11 = ________ H. 15 – 0 = ________

Word Problem

I. Hector gave his sister Carmen 5 packets of tomato seeds. Each packet of seeds has 6 seeds inside. How many tomato seeds did Hector give to Carmen?

1 packet of seeds

________ seeds

2 packets of seeds

________ seeds

3 packets of seeds

________ seeds

4 packets of seeds

________ seeds

5 packets of seeds

________ seeds

Math Problem:

Sentence: ________________________________

Answers: A. 9, B. 10, C. 13, D. 14, E. 19, F. 4, G. 11, H. 15, I. 1 packet of seeds: 6 seeds; 2 packets of seeds: 12 seeds; 3 packets of seeds: 18 seeds; 4 packets of seeds: 24 seeds; 5 packets of seeds: 30 seeds

Name: ______ Date: ______

66

Warm Up

Add or subtract.

A. 7 + 9 = ______	**C.** 16 + 2 = ______	**E.** 18 + 2 = ______	**G.** 9 + 8 = ______
B. 10 – 3 = ______	**D.** 10 – 4 = ______	**F.** 10 – 2 = ______	**H.** 19 – 9 = ______

Word Problem

I. Wally bought a bag of peanuts. There were 30 peanuts in the bag. Wally gave an elephant 2 peanuts a day as a small treat. How many days would it take to finish the bag of peanuts?

Math Problem:

Sentence: ______

Answers: A. 16, B. 7, C. 18, D. 6, E. 20, F. 8, G. 17, H. 10, I. 15 days

Name: ______________________ **Date:** ______________________

67

Warm Up

Write the number that comes in between.

A. 14, ________, 16 **C.** 81, ________, 83 **E.** 69, ________, 71 **G.** 32, ________, 34

B. 10, ________, 12 **D.** 70, ________, 72 **F.** 19, ________, 21 **H.** 39, ________, 41

Word Problem

I. An octopus has 8 arms. How many arms would 3 octopi have?

________ arms

________ arms

________ arms

Math Problem:

Sentence: ____

Answers: A. 15, B. 11, C. 82, D. 71, E. 70, F. 20, G. 33, H. 40, I. 24 arms

Warm Up

Name: Date:

68

Write the number that comes before.

A. ________, 25, 26

C. ________, 57, 58

E. ________, 8, 9

G. ________, 91, 92

B. ________, 39, 40

D. ________, 79, 80

F. ________, 14, 15

H. ________, 71, 72

Word Problem

I. Squirrels eat and gather acorns. Casey, the squirrel, gathered 8 acorns on Thursday. On Friday Casey gathered 14 acorns. On Saturday Casey gathered 13 acorns. How many acorns did Casey gather in 3 days? (Draw the number of acorns that Casey gathered on each day.)

Thursday:

Friday:

Saturday:

Math Problem:

Sentence: ________________________________

Answers: A. 24, B. 38, C. 56, D. 78, E. 7, F. 13, G. 90, H. 70, I. 35 acorns total—8 acorns on Thursday, 14 acorns on Friday, 13 acorns on Saturday.

Warm Up

Name: ____________________ Date: ____________________

69

Write the number that comes next.

A. 11, 12, ________

C. 87, 88, ________

E. 61, 62, ________

G. 41, 42, ________

B. 14, 15, ________

D. 90, 91, ________

F. 71, 72, ________

H. 3, 4, ________

Word Problem

I. There are 14 plums in a tree. There are some plums on the grass. There are 20 plums in all. How many plums are on the grass? Glue the correct number of plums on the grass, using the counters provided on page 254.

Math Problem:

Sentence: ____________________

Answers: A. 13, B. 16, C. 89, D. 92, E. 63, F. 73, G. 43, H. 5, I. 6 plums

Name: ______________________ Date: ______________________

70

Warm Up

Write the number that comes before.

A. ________, 41, 42

C. ________, 16, 17

E. ________, 80, 81

G. ________, 49, 50

B. ________, 9, 10

D. ________, 99, 100

F. ________, 61, 62

H. ________, 1, 2

Word Problem

I. There are 27 students in the class. If 13 of the students ate red strawberries, 7 students ate red apples, and 7 students ate pears, how many students ate red-colored fruit?

Math Problem:

Sentence: ____

Answers: A. 40, B. 8, C. 15, D. 98, E. 79, F. 60, G. 48, H. 0, I. 20 students

71

Name: **Date:**

Warm Up

Write the number that comes next.

A. 22, 23, ________ **C.** 39, 40, ________ **E.** 17, 18, ________ **G.** 48, 49, ________

B. 15, 16, ________ **D.** 44, 45, ________ **F.** 8, 9, ________ **H.** 0, 1, ________

Word Problem

I. There are 3 turtles. The first turtle laid 5 eggs. The second turtle laid 3 eggs. The third turtle laid 4 eggs. How many eggs did the three turtles lay in all?

turtle #1

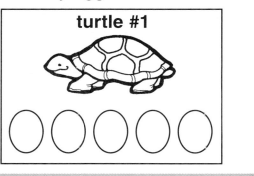

turtle #2

turtle #3

Math Problem:

Sentence: ____________________

Answers: A. 24, B. 17, C. 41, D. 46, E. 19, F. 10, G. 50, H. 2, I. 12 eggs

72

Warm Up

Name: ______________________ **Date:** ______________________

Add.

A.	**B.**	**C.**	**D.**
$13 + 5 = $ ___	$25 + 14 = $ ___	$7 + 12 = $ ___	$10 + 7 = $ ___

Word Problem

E. Megan has 12 crayons. She also has 8 markers. How many crayons and markers does Megan have? Glue the correct number of each item that Megan has by using the counters provided on page 254.

Math Problem:

Sentence: ______________________

Answers: A. 18, B. 39, C. 19, D. 17, E. 20 markers and crayons

Warm Up

Name: ____________________ Date: ____________________ **73**

Add.

A. $11 + 7 =$ ____

B. $14 + 4 =$ ____

C. $6 + 11 =$ ____

D. $10 + 9 =$ ____

Word Problem

E. There are 60 students going on a field trip to the local zoo. If a bus can carry 20 students, how many buses are needed to take the students to the zoo? Circle the correct number of buses needed?

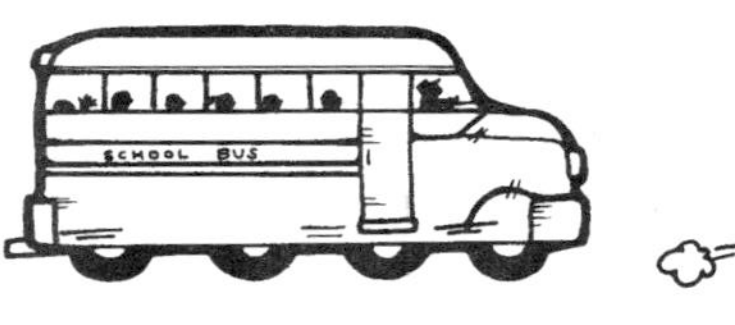

Math Problem:

Sentence: ____________________

Answers: A. 18, B. 18, C. 17, D. 19, E. 3 buses

Name: ____________________ **Date:** ____________________

74

Warm Up

Add.

A. 11 + 2 + 3 = ________

C. 2 + 12 + 2 = ________

B. 3 + 10 + 11 = ________

D. 11 + 14 + 23 = ________

Word Problem

E. Nina and Juan each bought a ticket for a baseball game. If each ticket costs $23, what was the total price for Nina and Juan's tickets?

Math Problem:

Sentence: ____

Answers: A. 16, B. 24, C. 16, D. 48, E. $46

Warm Up

Name: Date: **75**

Write the missing number.

A. 10 – ________ = 8 **C.** 16 – ________ = 0 **E.** 8 – ________ = 4 **G.** 27 – ________ = 4

B. 7 – ________ = 3 **D.** 29 – ________ = 1 **F.** 15 – ________ = 2 **H.** 13 – ________ = 3

Word Problem

I. The toy store has 22 basketballs for sale. If 8 girls and 7 boys each buy one basketball, how many basketballs will be left? (Draw the correct number of basketballs to fill in the chart below.)

girls	
boys	
balls remaining	

Math Problem:

Sentence: ________________________________

Answers: A. 2, B. 4, C. 16, D. 28, E. 4, F. 13, G. 23, H. 10, I. 7 basketballs remaining

76

Warm Up Name: ________ Date: ________

Add and subtract.

A. 10 – 1 + 0 = ________

C. 10 + 3 – 1 = ________

B. 17 – 4 + 5 = ________

D. 2 + 7 – 9 = ________

Word Problem

E. One beetle has 6 legs. Draw the correct number of legs on each beetle. How many legs are there on 3 beetles?

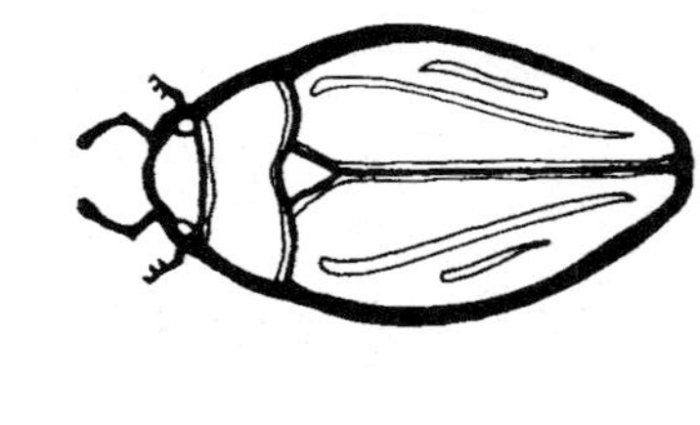
____

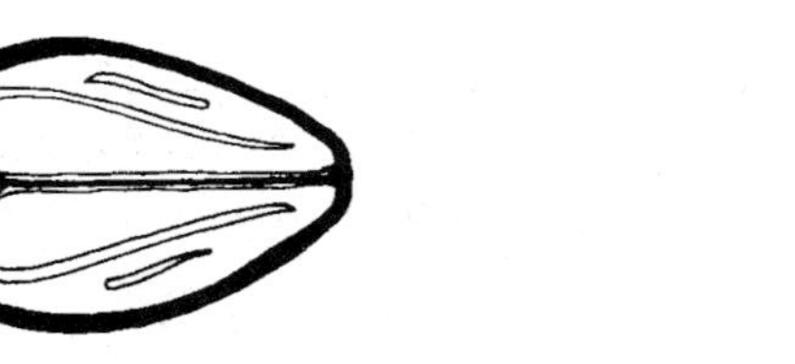
____

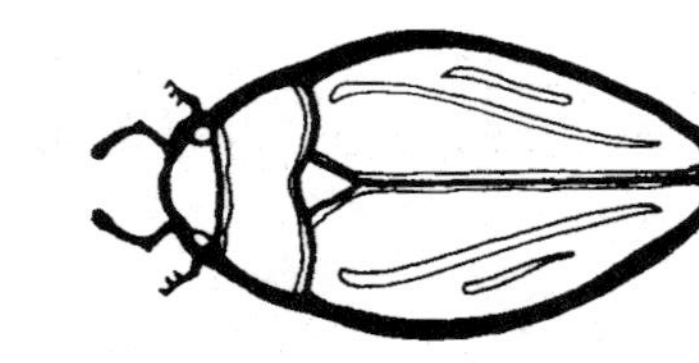
____

Math Problem:

Sentence: ________________________________

Answers: A. 9, B. 18, C. 12, D. 0, E. 18 legs

77

Warm Up

Name: ______________________ Date: ______________________

Add and subtract.

A. 19 – 5 + 3 = ________

B. 15 + 5 – 10 = ________

C. 26 + 3 – 2 = ________

D. 7 + 3 – 8 = ________

Word Problem

E. There are 10 different party hats at the store. Phillip picked out a special hat. Phillip picked a hat with an odd number. The hat's number is greater than 2 but less than 5. Which hat did Phillip pick?

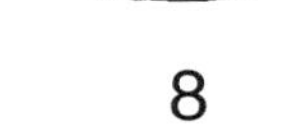

1 2 3 4 5 6 7 8 9 10

Math Problem:

Sentence: ____

Answers: A. 17, B. 10, C. 27, D. 2, E. 3

Warm Up

Name: ______________________ **Date:** ______________

78

Look at each pair of numbers. Use the > (greater than) or < (less than) sign. Complete each sentence.

A. 9 ◯ 17 **B.** 5 ◯ 7 **C.** 12 ◯ 2 **D.** 13 ◯ 19

_____ is less than _____. _____ is less than _____. _____ is greater than _____. _____ is less than _____.

Word Problem

E. Mrs. Wong wants to make brownies for all her students. There are 14 students in her class. If she wants to give each student 2 brownies, how many brownies does she have to bake? Glue the total number of brownies that Mrs. Wong needs to bake from the counters on page 254.

Math Problem:

Sentence: ____

Answers: A. <, B. <, C. >, D. <, E. 28 brownies

79

Name: ______________________ Date: ______________________

Warm Up

Add 6 to each number.

A. 10 + _______ = _______

B. 3 + _______ = _______

C. 6 + _______ = _______

D. 11 + _______ = _______

Word Problem

E. The Levin family orders pizzas for dinner. If each pizza is cut up into 8 slices, how many slices are there in 2 whole pizzas?

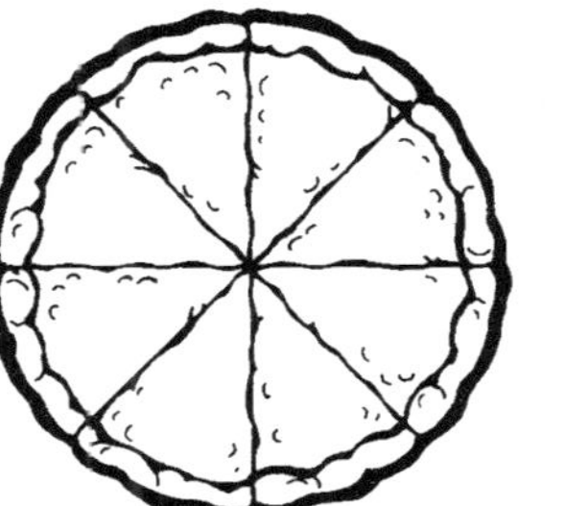

Math Problem:

Sentence: ____

Answers: A. 16, B. 9, C. 12, D. 17, E. 16 slices

Warm Up Name: Date: **80**

Fact Families: Use 4, 6, and 10 to make 2 addition problems and 2 subtraction problems.

A. 4 + ________ = ________

B. 6 + ________ = ________

C. ________ – ________ = 6

D. ________ – 6 = ________

Word Problem

E. Mr. Birdsong has 3 bird nests in his backyard. If there are 6 birds living in each nest, what is the total number of birds living in Mr. Birdsong's backyard?

Math Problem:

Sentence: ______________________________

Answers: A. 4 + 6 = 10, B. 6 + 4 = 10, C. 10 – 4 = 6, D. 10 – 6 = 4, E. 18 birds

81

Warm Up

Name: ______________________ Date: ______________________

Subtract.

A. $10 - 10 =$ ____

B. $8 - 0 =$ ____

C. $1 - 0 =$ ____

D. $6 - 1 =$ ____

Word Problem

E. Farmer Frank planted vegetables in his garden. Farmer Frank planted 4 rows of corn with 5 corn stalks in each row. How many corn stalks did Farmer frank plant in all? Glue the correct number of counters provided on page 254.

Row 1

Row 3

Row 2

Row 4

Math Problem:

Sentence: ______________________

Answers: A. 0, B. 8, C. 1, D. 5, E. 20 corn stalks

Warm Up

Name: **Date:** **82**

Subtract.

A.	**B.**	**C.**	**D.**
15	19	13	20
-13	-8	-3	-6

Word Problem

Mr. Franklin's class took a trip on a glass-bottom boat. The students were able to see different kinds of sea animals. Look at the graph below. It shows some of the different kinds of sea animals the students saw. What can you tell by looking at the graph? Write a math problem and a sentence about the graph.

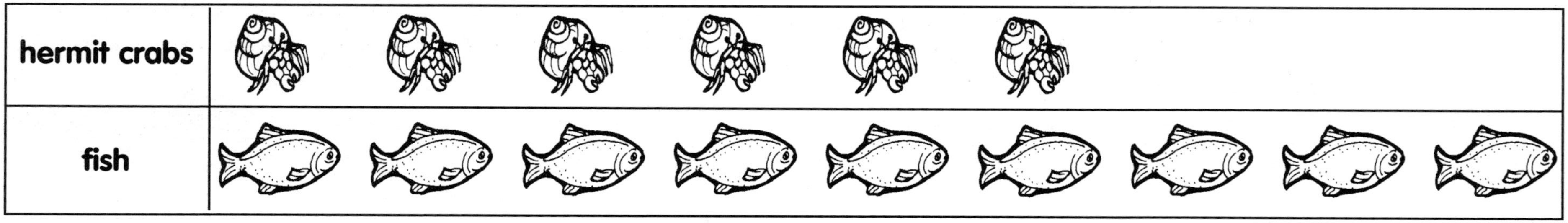

hermit crabs	
fish	

Math Problem:

Sentence: ____________________

Answers: A. 2, B. 11, C. 10, D. 14

83

Warm Up

Name: ____________ **Date:** ____________

Subtract.

A. $20 - 11$

B. $17 - 12$

C. $20 - 10$

D. $19 - 5$

Word Problem

E. Daniel has a lot of toys in his bedroom. If you add 5 to the mystery number of toys in Daniel's bedroom and then subtract 3, the answer will be 7. How many toys does Daniel have in his bedroom?

**0 1 2 3 4 5 6 7 8 9 10 11 12 13
14 15 16 17 18 19 20**

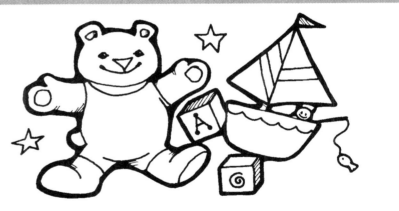

Math Problem:

Sentence: ____________

Answers: A. 9, B. 5, C. 10, D. 14, E. 5 toys

84

Warm Up

Name: ____________ Date: ____________

Subtract.

A.	B.	C.	D.
$17 - 11$	$19 - 6$	$12 - 11$	$10 - 4$

Word Problem

E. Gwen cut 25 flowers to put into the 5 flower pots. How many flowers does Gwen need to put into each pot so that each will have the same number of flowers? Draw the number of flowers that Gwen needs in each flower pot.

Math Problem:

Sentence: ____________

Answers: A. 6, B. 13, C. 1, D. 6, E. 5 flowers in each pot

85

Warm Up

Name: **Date:**

Subtract.

A. $\begin{array}{r} 12 \\ -\ 2 \\ \hline \end{array}$ **B.** $\begin{array}{r} 17 \\ -\ 10 \\ \hline \end{array}$ **C.** $\begin{array}{r} 29 \\ -\ 7 \\ \hline \end{array}$ **D.** $\begin{array}{r} 20 \\ -\ 12 \\ \hline \end{array}$

Word Problem

E. Garrett bought a board game for $17.95. Two days later he saw the same board game at another store for $19.50. How much did Garrett save?

Math Problem:

Sentence: ____________________

Answers: A. 10, B. 7, C. 22, D. 8, E. $1.55

Warm Up

Name: Date:

Subtract.

A.	B.	C.	D.
$10 - 7$	$10 - 6$	$10 - 5$	$17 - 4$

Word Problem

Vivian likes to collect seashells. She found 5 seashells on Saturday and 4 seashells on Sunday. How many seashells did she collect in all? Using the counters provided on page 255, glue the correct number of shells that Vivian found on Saturday and Sunday.

Saturday + Sunday = ☐

Math Problem:

Sentence: ____________________

Answers: A. 3, B. 4, C. 5, D. 13

Warm Up

Name: ______ Date: ______

87

Add.

A. $\begin{array}{r} 13 \\ +\ 15 \\ \hline \end{array}$

B. $\begin{array}{r} 14 \\ +\ 14 \\ \hline \end{array}$

C. $\begin{array}{r} 12 \\ +\ 18 \\ \hline \end{array}$

D. $\begin{array}{r} 20 \\ +\ 10 \\ \hline \end{array}$

Word Problem

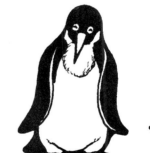

E. Penelope saw a whole bunch of penguins standing in line. There was 1 big penguin in the middle of the line. There were 5 small penguins on one side of the big penguin. There were 3 small penguins on the other sid eof the big penguin. How many penguins did Penelope see standing in a line?

Math Problem:

Sentence: ______

Answers: A. 28, B. 28, C. 30, D. 30, E. 9 penguins

88

Warm Up

Name: ____________ Date: ____________

Add or subtract.

A. 15 + 5 = ______ **C.** 16 – 6 = ______ **E.** 4 + 14 = ______ **G.** 20 – 7 = ______

B. 23 + 3 = ______ **D.** 17 – 5 = ______ **F.** 5 + 13 = ______ **H.** 2 + 26 = ______

Word Problem

I. Jose bought his mother flowers and a box of chocolates for her birthday. If he spent $9.95 for the flowers and $7.25 for the chocolates, how much money did Jose spend for the gifts?

Math Problem:

Sentence: ______________________________

Answers: A. 20, B. 26, C. 10, D. 12, E. 18, F. 18, G. 13, H. 28, I. $17.20

89

Name: ____________ Date: ____________

Warm Up

Number the months of the year in chronological order.

____ December	____ January	____ October	____ September
____ March	____ November	____ February	____ August
____ June	____ April	____ July	____ May

Word Problem

A. Annie's birthday is in January. Her mother's birthday is 6 months after her birthday. In what month was her mother born? ____________

B. Ethan's birthday is 3 months before August. In what month is Ethan's birthday? ____________

C. Wendy's birthday is in September. Her sister Julie's birthday is 6 months before Wendy's. In what month is Julie's birthday? ____________

Math Problem:

Sentence: ____

Answers: January, February, March, April, May, June, July, August, September, October, November, December; A. July, B. May, C. March

Warm Up

Name: Date:

90

Add.

A. 13 + 14 = ________ **C.** 11 + 3 = ________ **E.** 9 + 11 = ________ **G.** 12 + 16 = ________

B. 14 + 3 = ________ **D.** 10 + 17 = ________ **F.** 6 + 14 = ________ **H.** 9 + 28 = ________

Word Problem

Yolanda asked her friends, "Which vegetable do you like best?" Yolanda made a graph to show which vegetable her friends liked best. Write a math problem and a sentence about Yolanda's graph.

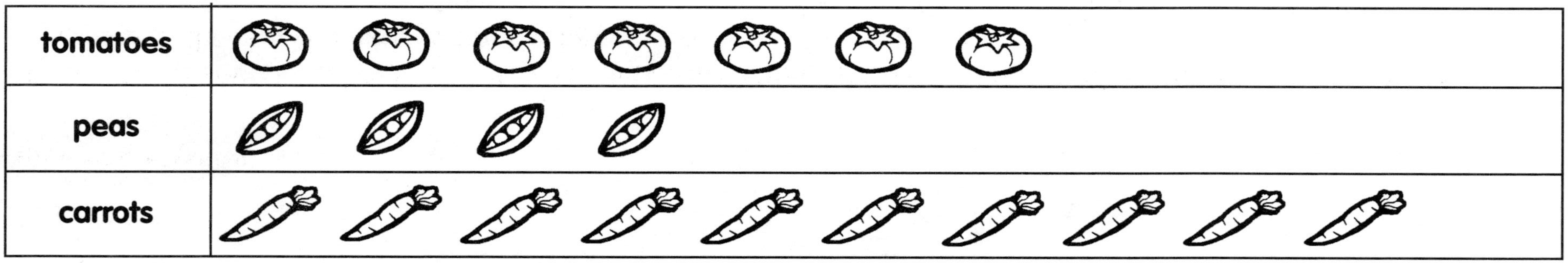

tomatoes	
peas	
carrots	

Math Problem:

Sentence: ____

Answers: A. 27, B. 17, C. 14, D. 27, E. 20, F. 20, G. 28, H. 37

Warm Up

Name: **Date:** **91**

Add and subtract.

A. 21 + 8 = ________ **C.** 9 + 10 = ________ **E.** 29 – 13 = ________ **G.** 29 – 5 = ________

B. 8 + 21 = ________ **D.** 20 + 9 = ________ **F.** 19 – 6 = ________ **H.** 27 – 14 = ________

Word Problem

I. There are several animals on the Johnson farm. There are pigs, horses, chickens, ducks, and sheep. Which of these animals only have 2 legs?

Math Problem:

Sentence: ____________________

Answers: A. 29, B. 29, C. 19, D. 29, E. 16, F. 13, G. 24, H. 13, I. chickens and ducks

Name: ____________ **Date:** ____________

92

Warm Up

Add or subtract.

A. 13 + 4 = ______ **C.** 8 + 11 = ______ **E.** 12 + 3 = ______ **G.** 10 + 19 = ______

B. 19 – 6 = ______ **D.** 19 – 11 = ______ **F.** 29 – 18 = ______ **H.** 29 – 20 = ______

Word Problem

I. Charles and Randy had gathered a total of 15 marbles. Charles found an even number of marbles. Randy found an odd number of marbles. Charles found one more marble than Randy. How many marbles did Charles and Randy find? Fill in the chart to find the solution.

Charles		Randy	
______	+	______	= 15 marbles
______	+	______	= 15 marbles
______	+	______	= 15 marbles
______	+	______	= 15 marbles
______	+	______	= 15 marbles
______	+	______	= 15 marbles

Math Problem:

Sentence: ______________________________

Answers: A. 17, B. 13, C. 19, D. 8, E. 15, F. 11, G. 29, H. 9, I. Charles found 8 marbles, and Randy found 7 marbles.

Warm Up

Name: Date:

A. Count by 3's. Write the numbers on the lines. The first one has been done for you.

3 ______ ______ ______ ______ ______ ______

Word Problem

B. The Gupta family is putting a jigsaw puzzle together. So far they have used 35 pieces of the 50-piece puzzle. How many more pieces need to be placed to finish the puzzle?

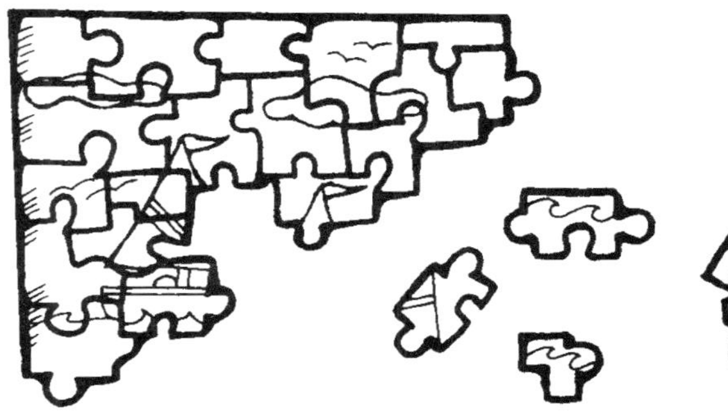

Math Problem:

Sentence: ______________________________

Answers: A. 3, 6, 9, 12, 15, 18 B. 15 pieces

94

Warm Up

Name: Date:

Add.

A. 14 + 13 = ________

B. 13 + 24 = ________

C. 12 + 12 = ________

D. 12 + 0 = ________

E. 15 + 5 = ________

F. 11 + 11 = ________

G. 14 + 16 = ________

H. 16 + 4 = ________

Word Problem

I. There are 20 different movies showing at the local cineplex. If a person wants to see all of the movies showing over the span of 10 days, how many movies does he or she need to see per day?

Math Problem:

Sentence: ________________________________

Answers: A. 27, B. 37, C. 24, D. 12, E. 20, F. 22, G. 30, H. 20, I. 2 movies

95

Warm Up Name: ____________ Date: ____________

Answer the question below each picture.

A. How many ears? ______

B. How many legs? ______

C. How many noses? ______

D. How many eyes? ______

Word Problem

E. Henry saw 5 lions. If each lion has 2 eyes and 1 tail, how many eyes and how many tails do the 5 lions have?

1 lion	2 lions	3 lions	4 lions	5 lions
______ eyes	______ eyes	______ eyes	______ eyes	______ eyes
______ tails	______ tails	______ tails	______ tails	______ tails

Math Problem:

Sentence: ____

Answers: A. 4, B. 4, C. 3, D. 4, E. 10 eyes and 5 tails

Warm Up **Name:** ____________ **Date:** ____________

96

Solve each riddle.

A. I am a shape. I do not have any corners. What am I? ____________

B. I am a shape. I have 3 corners. What am I? ____________

C. I am a shape. I have 4 square corners and 4 equal sides. What am I? ____________

D. I am a shape. I have 2 long sides, short sides, and 4 square corners. What am I? ____________

Word Problem

Identify the different shapes in the picture below. Write a math problem about the picture and what you see in it.

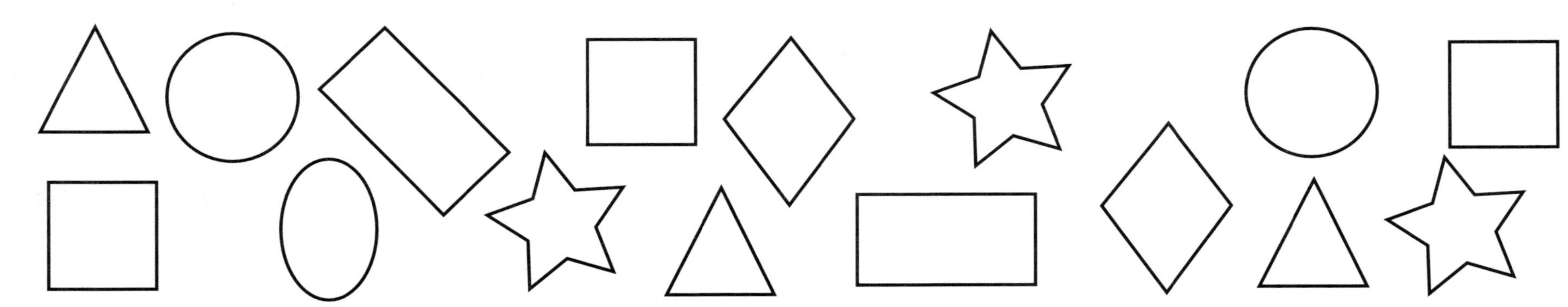

Math Problem:

Sentence: ____________

Answers: A. circle, B. triangle, C. square, D. rectangle

Warm Up

Name: **Date:**

97

Trace the numbers 11–20.

11 12 13 14 15 16 17 18 19 20

Word Problem

Many ants live in a colony. There are 40 ants in this colony. If the ants were to be divided into groups of 10, how many groups of 10 would the ants of this colony make?

Math Problem:

Sentence: ____________________

Answer: There would be 4 groups of ten ants.

Warm Up

Name: **Date:** **98**

Write the numbers 11 to 20.

Word Problem

Miguel and his friends decided to go fishing at the local lake. On Wednesday, Miguel caught 4 fish. On Thursday, Eric caught 3 fish. On Friday, Eva caught 5 fish. How many fish did Miguel and his friends catch? (Draw the number of fish that each person caught during the 3-day fishing trip.)

Wednesday	Thursday	Friday

Math Problem:

Sentence: ____________________

Answer: Miguel and his friends caught a total of 12 fish.

Warm Up

Name: ______________________ Date: ______________________

99

Read each number word. Trace each number word.

eleven twelve thirteen

Word Problem

For the Perez family, there are 3 birthdays to be celebrated during this week. Mr. Perez's birthday is on Thursday. Penelope's birthday is two days before her father's birthday. Eduardo's birthday is the day before Penelope's birthday. On what days are Penelope's and Eduardo's birthdays? Circle the day of the week that is Penelope's birthday. Draw a rectangle around the day of the week that is Eduardo's birthday.

Sunday	Thursday
Monday	Friday
Tuesday	Saturday
Wednesday	

Math Problem:

Sentence: ______________________

Answer: Penelope's birthday is on Tuesday. Eduardo's birthday is on Monday.

Warm Up

Name: **Date:**

Read each number word. Trace each number word.

fourteen fifteen sixteen

Word Problem

Julian had 6 puppies and 5 kittens. He gave Ross 3 puppies and 2 kittens. How many puppies and kittens does Julian have left? Cut out the counters provided on page 255 and glue the correct number of puppies and kittens that each person has.

Julian

Ross

Math Problem:

Sentence: ____________________

Answer: Julian has 3 puppies and 3 kittens. Ross has 3 puppies and 2 kittens.

Warm Up

Name: Date: **101**

Read each number word. Trace each number word.

seventeen eighteen

nineteen twenty

Word Problem

Jenny planted a vegetable garden on her farmland. She planted 3 rows of cabbages, 5 rows of carrots, 2 rows of spinach, and 4 rows of squash. How many rows did she plant in all?

Math Problem:

Sentence: ______________________

Answer: Jenny planted 14 rows of vegetables.

102

Name: **Date:**

Warm Up

Write the numbers 11 to 20.

Word Problem

A stamp is 1 inch (2.54 cm) wide. If you line up 5 stamps, what is the combined width of these stamps?

Math Problem:

Sentence: ____________________

Answer: The 5 stamps have a combined width of 5" (13 cm).

103

Name: **Date:**

Warm Up

Look at the hats. Answer the questions.

A. How are the hats alike?	**B.** How are the hats different?

Word Problem

C. How many top hats are in only the square? _____

D. How man top hats are there in both the square and the circle? _____

E. How many caps are there in all? _____

F. How many caps are there in the square? _____

G. How many caps are in the circle? _____

H. Which hat is in the circle only, the square only, and both the circle and the square? _____

Math Problem:

Sentence: ____________________

Answers: A. They are all baseball caps. B. None of the hats are the same because they are differnet shapes and styles. C. 2, D. 2, E. 5, F. 2, G. 3. H. bonnet

Warm Up

Name: Date: **104**

Look at each number given. Write a number that is 2 more.

A. 10 __________ **C.** 6 __________ **E.** 12 __________

B. 5 __________ **D.** 0 __________ **F.** 8 __________

Word Problem

G. Mindy had 12 marbles. Her sister, Margo, gave her 5 more marbles. Her brother, Marco, gave her some marbles too. Now Mindy has 22 marbles. How many marbles did her brother give her? Color the correct number of marbles that Marco gave Mindy.

Math Problem:

Sentence: ______________________________

Answers: A. 12, B. 7, C. 8, D. 2, E. 14, F. 10, G. 5 marbles

Warm Up

Name: ________ **Date:** ________

Look at each number given. Write a number that is 2 less.

A. 4 ________ **C.** 2 ________ **E.** 15 ________

B. 7 ________ **D.** 9 ________ **F.** 12 ________

Word Problem

G. If an apple has 6 seeds, how many seeds are there in 4 apples?

1 apple 2 apples 3 apples 4 apples

______ seeds ______ seeds ______ seeds ______ seeds

Math Problem:

Sentence: ________________________

Answers: A. 2, B. 5, C. 0, D. 7, E. 13, F. 10, G. 24 seeds

106

Name: **Date:**

Warm Up

Look at each number sequence. Write the missing numbers in each sequence.

A. _______, 10, 11, _______

B. 22, _______, _______, 25

C. 35, _______, _______

D. _______, 18, 19, _______

Word Problem

E. The Ryan family decided to plant some trees in their backyard. They planted the following trees:

- The oak tree was planted first.
- The apple tree was planted last.
- The maple tree was planted next to the apple tree.
- The pine tree was planted next to the oak tree.

Write these trees in the order that they were planted.

1. _______________
2. _______________
3. _______________
4. _______________

Math Problem:

Sentence: _______________________________

Answers: A. 9, 12, B. 23, 24, C. 36, 37, D. 17, 20, E. oak tree, pine tree, maple tree, and apple tree

Warm Up

Name: Date:

107

Read each number word. Complete each sequence by writing the number word that precedes or follows the number word given.

A. thirteen, ____________, fifteen

B. fourteen, ____________, sixteen

C. nine, ten, ____________

D. seventeen, eighteen, ____________

Word Problem

E. The teacher assigned science experiments that were to be presented in class. Each student had to work with a partner. If there were 24 students in the classrooms, how many experiments were assigned?

Math Problem:

Sentence: ____

Answers: A. fourteen, B. fifteen, C. eleven, D. nineteen, E. 12 experiments

Name: **Date:** **108**

Warm Up

Read each number word. Write the number on the line.

nineteen ________ sixteen ________ fourteen ________

twenty ________ thirteen ________ eighteen ________

Word Problem

There were 18 acorns on the ground. A hungry squirrel came along and took 8 of the acorns. How many acorns are still on the ground? Cross out the 8 acorns that the squirrel took and count how many still remain.

Math Problem:

Sentence: ________________________

Answer: 19, 20, 16, 13, 14, 18; 10 acorns

Warm Up

Name: ____________ Date: ____________

109

Read each number. Write the number on the line.

11 ______________ 13 ______________ 17 ______________

12 ______________ 14 ______________ 19 ______________

Word Problem

An hour before the circus put on its first performance, 25 people were already waiting inside the circus tent. A half hour later, 13 more people went inside the tent. How many people were inside the circus tent to watch the first performance?

Math Problem:

Sentence: ______________________________

Answer: eleven, twelve, thirteen, fourteen, seventeen, nineteen; 38 people

Warm Up **Name:** **Date:** **110**

Read each number. Write the number word on the line.

16 ______ 19 ______ 7 ______

20 ______ 14 ______ 10 ______

Word Problem

Yolanda has a grocery bag full of groceries. In the bag, there are 6 lemons, 13 cherries, 5 tomatoes, and 3 cucumbers. How many pieces of red fruit and/or vegetables does she have in the bag?

Math Problem:

Sentence: ______

Answer: sixteen, twenty, nineteen, fourteen, seven, ten; 18 red fruits/vegetables

Warm Up

Name: ______ **Date:** ______ **111**

Read each problem and write the answer on the line provided.

How old are you right now? ______

How old will you be in 10 years? ______

How old will you be in 5 years? ______

How old will you be in 20 years? ______

Word Problem

Aunt Rosa and her niece and nephew like to go to the library. Last weekend they borrowed a lot of books. Aunt Rosa borrowed 24 books. Her nephew Raphael borrowed 6 books. Her niece Charlene borrowed 3 books. What is the total number of books that were borrowed? Using the counters on page 255, glue the correct number of books for each person.

Raphael **Aunt Rosa** **Charlene**

Math Problem:

Sentence: ______

Answer: The total number of books borrowed was 11.

112

Warm Up

Name: **Date:**

Write the number that comes in between.

A. 11, ______, 13

B. 20, ______, 22

C. 89, ______, 91

D. 13, ______, 15

E. 98, ______, 100

F. 62, ______, 64

G. 51, ______, 53

H. 40, ______, 42

Word Problem

I. Each clown is juggling 4 balls. If there are 4 clowns juggling at the same time, how many balls are being juggled at one time? (Draw the number of balls that the other three clowns are juggling.)

Math Problem:

Sentence: ______________________________

Answers: A. 12, B. 21, C. 90, D. 14, E. 99, F. 63, G. 52, H. 41, I. 16 balls

113

Warm Up

Name: Date:

Write the missing numbers.

A. 11, ________, 13, ________, 15

B. 12, ________, ________, 15, 16

C. 14, ________, 16, ________, ________

D. 16, ________, 18, ________, ________

Word Problem

Measure the width of each shape.

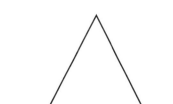

E. How wide is the circle? ________

F. How wide is the triangle? ________

G. How wide is the square? ________

H. How wide is the rectangle ________?

Math Problem:

Sentence: ____

Answers: A. 12, 14, B. 13, 14, C. 15, 17, 18, D. 17, 19, 20 E. 1 inch (2.54 cm), F. 1 inch (2.54 cm), G. 2 inches (5 cm), H. 2 inches (5 cm)

Warm Up

Name: **Date:**

114

Look at each pair of numbers. Use the > or < sign. Complete each sentence.

A. 7 ◯ 11 — ______ is less than ______.

B. 9 ◯ 14 — ______ is less than ______.

C. 4 ◯ 1 — _____ is greater than _____.

D. 20 ◯ 6 — _____ is greater than _____.

Word Problem

E. There were 18 kids riding the school bus home. At the first stop, the bus driver let 4 kids off of the bus. At the second stop, 5 more kids got off the bus. At the third stop, the remaining kids got off the bus. How many kids got off at the third stop?

Math Problem:

Sentence: ____________________

Answers: A. 7 < 11, B. 9 < 14, C. 4 > 1, D. 20 > 6, E. 9 kids

Name: Date: **115**

Warm Up

Write the missing number.

A. 15 – ______ = 12 **C.** 26 – ______ = 14 **E.** 12 – ______ = 5 **G.** 11 – ______ = 7

B. 17 – ______ = 4 **D.** 14 – ______ = 11 **F.** 19 – ______ = 8 **H.** 10 – ______ = 3

Word Problem

I. Serena has 2 dogs. Each dog eats 5 dog biscuits a day. How many dog biscuits in all do Serena's dogs eat each day? Glue the correct number of dog biscuit counters provided on page 255 to show your answer.

dog #1	dog #2

Math Problem:

Sentence: ______________________________

Answers: A. 3, B. 13, C. 12, D. 3, E. 7, F. 11, G. 4, H. 7, I. 10 dog biscuits

116

Warm Up

Name: Date:

Add.

A.	B.	C.	D.
10 + 12	9 + 10	8 + 12	18 + 10

Word Problem

Cut out the buttons on page 255. How many different ways can you sort the buttons? Sort the button counters and glue them below. Write a math problem and a sentence telling about how you sorted the buttons.

Math Problem:

Sentence: ____________________

Answers: A. 22, B. 19, C. 20, D. 28

117

Warm Up

Name: Date:

Make 2 addition and 2 subtraction problems that have 11 as the answer.

________ + ________ = ________   ________ – ________ = ________

________ + ________ = ________   ________ – ________ = ________

Word Problem

Count the number of each fruit or vegetable in each row. Write a math problem and sentence about what you observe in the chart below.

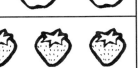

red apples	
strawberries	

potatoes	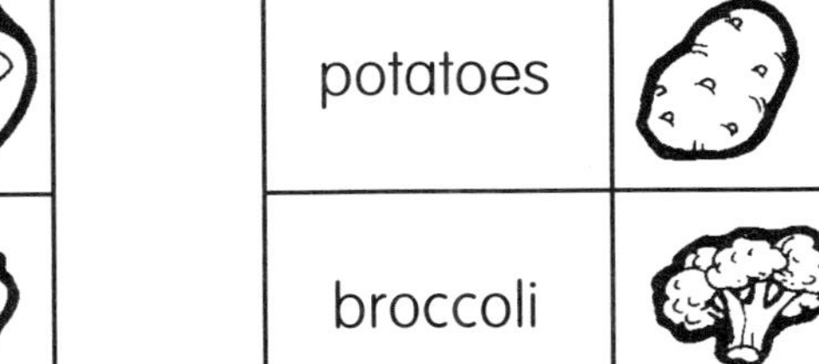
broccoli	

Math Problem:

Sentence: ______________________________

Warm Up

Name: **Date:** **118**

Add or subtract even numbers.

A. $10 + 12$

B. $18 - 6$

C. $10 - 4$

D. $14 + 4$

Word Problem

Using only even numbers in your problem, write your own word problem.

Math Problem:

Sentence: ____________________

Answers: A. 22, B. 12, C. 6, D. 18

119

Name: Date:

Warm Up

Add or subtract odd numbers.

A. $\begin{array}{r} 11 \\ -\ 7 \\ \hline \end{array}$

B. $\begin{array}{r} 15 \\ -\ 9 \\ \hline \end{array}$

C. $\begin{array}{r} 25 \\ -\ 3 \\ \hline \end{array}$

D. $\begin{array}{r} 29 \\ +\ 7 \\ \hline \end{array}$

Word Problem

E. Froggy is a great jumper. Froggy jumped from lily pad to lily pad to get from one side of the pond to the other. Each lily pad was 5 feet away from the next lily pad. If Froggy jumped 30 feet in all, on how many lily pads did he jump to get from one side of the pond to the other?

Math Problem:

Sentence: ______________________________

Answers: A. 4, B. 6, C. 22, D. 36, E. 5 lily pads

120

Warm Up

Name: Date:

Use tally marks to show each number.

14

17

15

18

Word Problem

Look at 1 die. Draw the number of pips you see on each side of the die. Count the total number of pips you made. What is the total number of pips on 1 die?

Math Problem:

Sentence: ______________________________

Answer: There is a total of 21 pips on 1 die.

Warm Up

Name: **Date:** **121**

Write the missing number.

A. ______ + 2 = 15 **C.** ______ + 4 = 10 **E.** ______ + 10 = 19 **G.** ______ + 10 = 15

B. ______ + 2 = 17 **D.** ______ + 11 = 18 **F.** ______ + 6 = 12 **H.** ______ + 1 = 11

Word Problem

I. Deanna loves to grow sunflowers. Last year, Deanna grew 15 sunflowers. This year, Deanna grew 12 sunflowers. How many more sunflowers did Deanna grow last year than this year?

last year

this year

Math Problem:

Sentence: ____

Answers: A. 13, B. 15, C. 6, D. 7, E. 9, F. 6, G. 5, H. 10, I. 3 sunflowers

Name: Date:

122

Warm Up

Write the missing number.

A. 9 + ________ = 11 **C.** 6 + ________ = 9 **E.** 5 + ________ = 12 **G.** 7 + ________ = 11

B. 7 + ________ = 8 **D.** 5 + ________ = 10 **F.** 8 + ________ = 9 **H.** 3 + ________ = 10

Word Problem

I. How many ears and tails are on all the mice inside the circle? (*Hint:* Each mouse has a pair of ears and a tail.)

number of mice	number of ears	number of tails
1	________	________
2	________	________
3	________	________
4	________	________
5	________	________

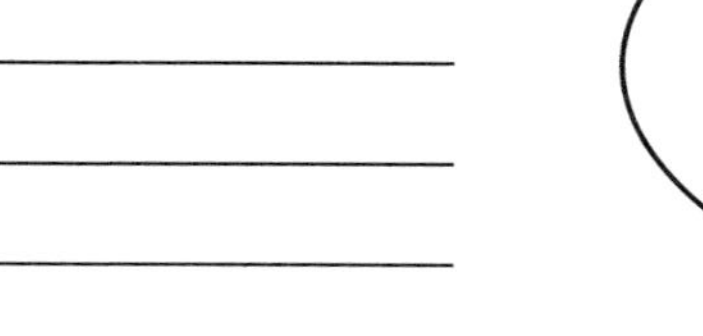
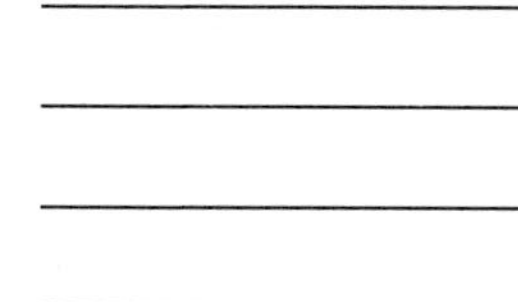

Math Problem:

Sentence: ____

Answers: A. 2, B. 1, C. 3, D. 5, E. 7, F. 1, G. 4, H. 7, I. 10 ears, 5 tails

Warm Up

Name: Date:

Add or subtract.

A. 10 + 7 = ________ **C.** 13 + 16 = ________ **E.** 17 – 7 = ________ **G.** 10 + 9 = ________

B. 14 – 12 = ________ **D.** 18 – 5 = ________ **F.** 19 – 14 = ________ **H.** 12 – 9 = ________

Word Problem

I. Every day Amanda goes for a walk. On Monday she picked 8 flowers. On Tuesday she picked 13 flowers. How many flowers did Amanda pick in all? Draw all of the flowers that Amanda picked.

Monday:

Tuesday:

Math Problem:

Sentence: ______________________________

Answers: A. 17, B. 2, C. 29, D. 13, E. 10, F. 5, G. 19, H. 3, I. 21 flowers

124

Warm Up

Name: Date:

Use the numbers 4, 9, and 13 to make 2 addition problems and 2 subtraction problems.

A. 9 + _______ = _______

B. 4 + _______ = _______

C. 13 – _______ = _______

D. _______ – 9 = _______

Word Problem

E. Mike wants to throw a surprise birthday party for his sister Lisa. He mailed 6 invitations on both Wednesday and Thursday. He mailed 9 invitations on Friday. How many people did Mike invite to Lisa's surprise birthday party? For each invitation mailed, color in a box for each day.

Wednesday	1	2	3	4	5	6	7	8	9	10
Thursday	1	2	3	4	5	6	7	8	9	10
Friday	1	2	3	4	5	6	7	8	9	10

Math Problem:

Sentence: _______________

Answers: A. 9 + 4 = 13, B. 4 + 9 = 13, C. 13 – 4 = 9, D. 13 – 9 = 4, E. 21 people

Warm Up

Name: **Date:** **125**

Subtract.

A. 14 – 8 = ________ **C.** 22 – 10 = ________ **E.** 14 – 13 = ________ **G.** 23 – 10 = ________

B. 23 – 11 = ________ **D.** 16 – 0 = ________ **F.** 10 – 8 = ________ **H.** 12 – 8 = ________

Word Problem

I. Mrs. King bought muffins and donuts for the students in her class. If she bought a dozen donuts and a dozen muffins, how many donuts and muffins did Mrs. King buy? Color the correct number of donuts and muffins that Mrs. King bought. (Remember: 1 dozen = 12)

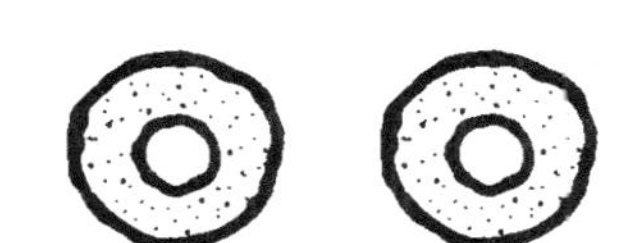

Math Problem:

Sentence: ______________________________

Answers: A. 6, B. 12, C. 12, D. 16, E. 1, F. 2, G. 13, H. 4, I. 24 donuts and muffins

Name: **Date:** **126**

Warm Up

Subtract.

A. 12 – 12 = ________ **C.** 14 – 6 = ________ **E.** 23 – 2 = ________ **G.** 11 – 3 = ________

B. 24 – 3 = ________ **D.** 19 – 10 = ________ **F.** 19 – 12 = ________ **H.** 27 – 7 = ________

Word Problem

I. A campground has 8 tents. If there are 24 people camping during the weekend, how many people can sleep in each tent so there is an equal number of people in each tent?

________ ________ ________ ________ ________ ________ ________ ________

Math Problem:

Sentence: ____

Answers: A. 0, B. 21, C. 8, D. 9, E. 21, F. 7, G. 8, H. 20, I. 3 people per tent

Name: Date: **127**

Warm Up

Add.

A. 12 + 12 = ________ **C.** 12 + 22 = ________ **E.** 12 + 10 = ________ **G.** 14 + 20 = ________

B. 17 + 12 = ________ **D.** 18 + 16 = ________ **F.** 19 + 14 = ________ **H.** 22 + 17 = ________

Word Problem

I. There are 2 jars and 16 jellybeans. How many different ways can you put the jellybeans into 2 jars?

jar #1		jar #2	
________	+	________	= 16 jellybeans
________	+	________	= 16 jellybeans
________	+	________	= 16 jellybeans
________	+	________	= 16 jellybeans
________	+	________	= 16 jellybeans

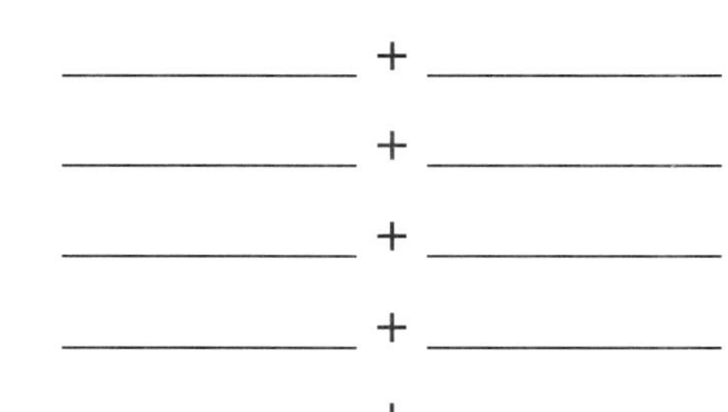

Math Problem:

Sentence: ____

Answers: A. 24, B. 29, C. 34, D. 34, E. 22, F. 33, G. 34, H. 39, I. Answers will vary.

128

Name: ____________ **Date:** ____________

Warm Up

Add.

A. 16 + 10 + 6 = ________

B. 17 + 3 + 11 = ________

C. 10 + 12 + 3 = ________

D. 19 + 2 + 10 = ________

Word Problem

E. As of Tuesday Davey had made 5 pies for a party. He needs 14 pies in all by Saturday. How many more pies does Davey need to make? Using the counters provided on page 256, glue the correct number of pies that Davey still needs to make.

Math Problem:

Sentence: ____

Answers: A. 32, B. 31, C. 25, D. 31, E. 9 pies

129

Warm Up

Name: Date:

Add.

A. 17 + 2 + 1 = ________

B. 13 + 2 + 6 = ________

C. 8 + 2 + 13 = ________

D. 7 + 3 + 14 = ________

Word Problem

E. Count the number of shapes in each butterfly's wings. Write the number of different shapes next to each butterfly.

________ circles

________ triangles

________ squares

________ circles

________ triangles

________ squares

Math Problem:

Sentence: ________________________________

Answers: A. 20, B. 21, C. 23, D. 24, E. Butterfly # 1—6 circles, 2 triangles, 2 squares; Butterfly # 2—4 circles, 4 triangles, 6 squares

130

Warm Up

Name: Date:

Subtract.

A. 12 – 4 = ________

C. 23 – 17 = ________

E. 15 – 11 = ________

G. 14 – 8 = ________

B. 21 – 6 = ________

D. 21 – 13 = ________

F. 19 – 10 = ________

H. 21 – 10 = ________

Word Problem

I. Grandma and Grandpa walk on the beach every day. Last week, Grandma and Grandpa found a lot of sand dollars. On Monday, they found 7 sand dollars. On Tuesday, they found 8 sand dollars. On Wednesday, they found 10 more sand dollars. How many sand dollars did Grandma and Grandpa find in all?

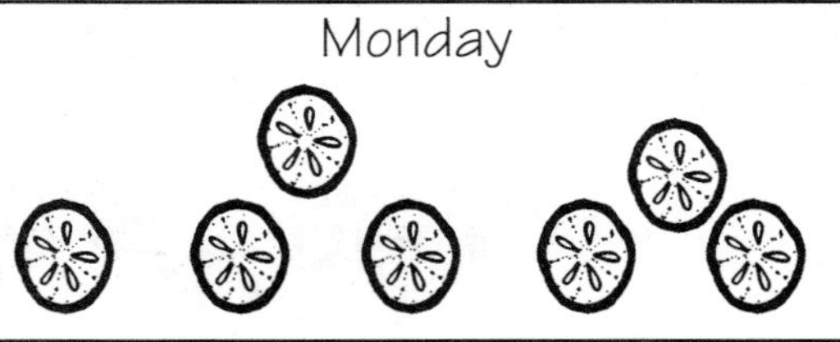

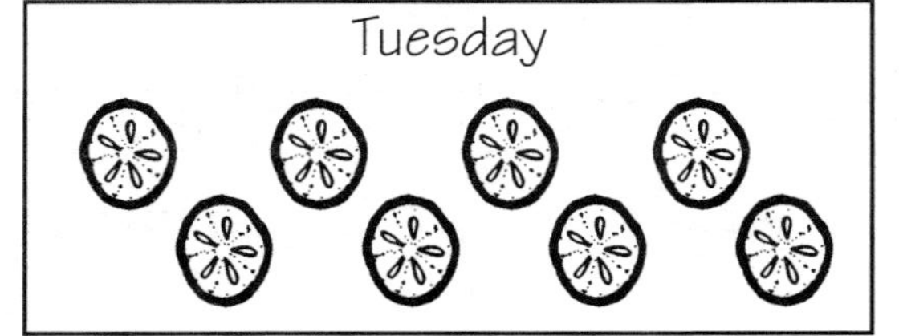

Math Problem:

Sentence: ________________________________

Answers: A. 8, B. 15, C. 6, D. 8, E. 4, F. 9, G. 6, H. 11, I. 25 sand dollars

Name: Date: **131**

Warm Up

Add.

A. 14 + 20 = ________ **C.** 13 + 15 = ________ **E.** 12 + 20 = ________ **G.** 15 + 16 = ________

B. 13 + 23 = ________ **D.** 14 + 19 = ________ **F.** 22 + 21 = ________ **H.** 10 + 21 = ________

Word Problem

I. The wind blew the leaves off of the tree. Julia picked up 20 leaves and 9 of the leaves were yellow. The rest of the leaves were red. How many red leaves did Julia pick up? Color the correct number of red and yellow leaves.

Math Problem:

Sentence: ________________________________

Answers: A. 34, B. 36, C. 28, D. 33, E. 32, F. 43, G. 31, H. 31, I. 11 red leaves

132

Name: ______________________ **Date:** ______________________

Warm Up

Look at each number and write a number that is 5 more.

A. 12 ________ **C.** 7 ________ **E.** 20 ________

B. 8 ________ **D.** 11 ________ **F.** 17 ________

Word Problem

G. Grace was reading a mystery book. She read 45 pages. If there are 96 pages in the book, how many more pages does Grace need to read to finish her book?

Math Problem:

Sentence: ____

Answers: A. 17, B. 13, C. 12, D. 16, E. 25, F. 22, G. 51 pages

Warm Up Name: Date: **133**

Look at each number and write a number that is 5 less.

A. 7 ________ **C.** 9 ________ **E.** 15 ________

B. 13 ________ **D.** 14 ________ **F.** 20 ________

Word Problem

G. A paperboy delivers 30 newspapers every morning. If it takes him 2 hours to deliver all the papers, how many papers does he deliver every hour?

Math Problem:

Sentence: ________________________________

Answers: A. 2, B. 8, C. 4, D. 9, E. 10, F. 15, G. 15 newspapers per hour

Warm Up

Name: Date: **134**

Circle the one that holds more. Circle the one that holds less.

A.

B.

C.

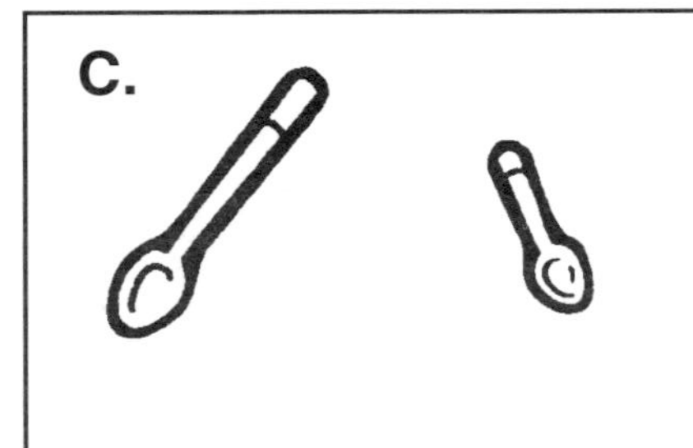

D.

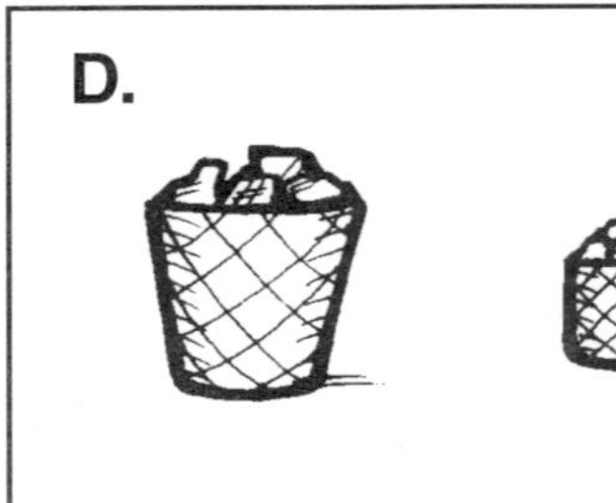

Word Problem

E. Jamie wants to send a letter to her grandparents. She needs to put a stamp on the envelope. To mail a regular first-class letter, a 33¢ stamp is needed. If she has saved 25¢ to buy a stamp, how much more does she need to save to buy a stamp to put on her letter?

Circle the correct combination of coins that Jamie still needs to save.

Math Problem:

Sentence: ____________________

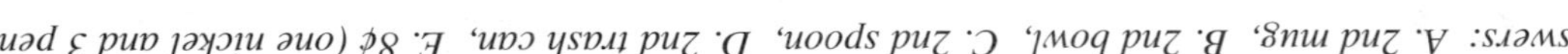

Answers: A. 2nd mug, B. 2nd bowl, C. 2nd spoon, D. 2nd trash can, E. 8¢ (one nickel and 3 pennies)

Warm Up

Name: ______________________ Date: ______________________

Circle the month that comes first.

A. January March

B. December July

C. February November

D. June May

Word Problem

The Martin family has 5 birthdays this month!

- Mr. Martin's birthday is on the 7th.
- Mrs. Martin's birthday is 3 days after Mr. Martin's.
- Morris Martin's birthday is 2 days after his mom's.
- Marcie Martin's birthday is 2 days after her brother's.
- Millie Martin's birthday is 6 days after her sister's.

What is the date of each family member's birthday?

E. Mrs. Martin ________.

F. Morris Martin ________.

G. Marcie Martin ________.

H. Millie Martin ________.

May						
S	M	T	W	Th	F	S
	1	2	3	4	5	6
7	8	9	10	11	12	13
14	15	16	17	18	19	20
21	22	23	24	25	26	27
28	29	30	31			

Math Problem:

Sentence: ______________________

Answers: A. January, B. July, C. February, D. May, E. May 10, F. May 12, G. May 14, H. May 20

Warm Up

Name: **Date:**

136

Look at the time on each clock. Write the correct time.

A.  ____ : ____

B. 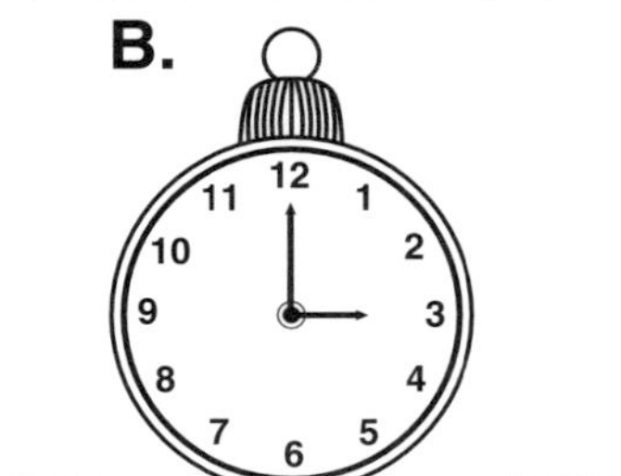 ____ : ____

C. ____ : ____

D. 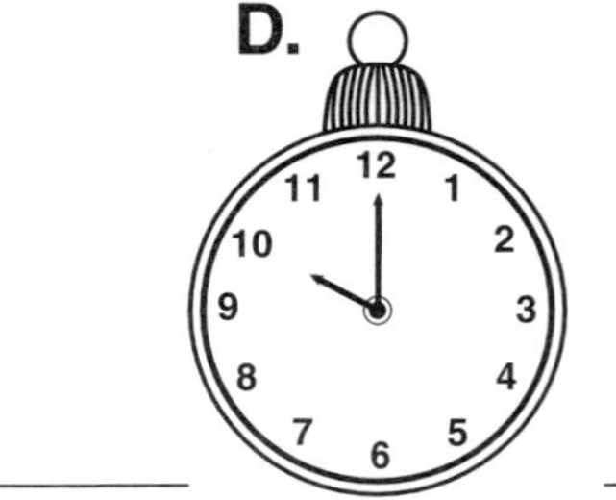 ____ : ____

Word Problem

E. Paula went to the park to play with some friends. She played for 4 hours. She stopped playing at 6:00 P.M. What time did Paula start playing at the park?

Math Problem:

Sentence: ________________________________

Answers: A. 8:00, B. 3:00, C. 6:00, D. 10:00, E. 2:00 P.M.

137

Warm Up

Name: ______________________ Date: ______________________

Look at the time on the clock. Write the correct time.

A. 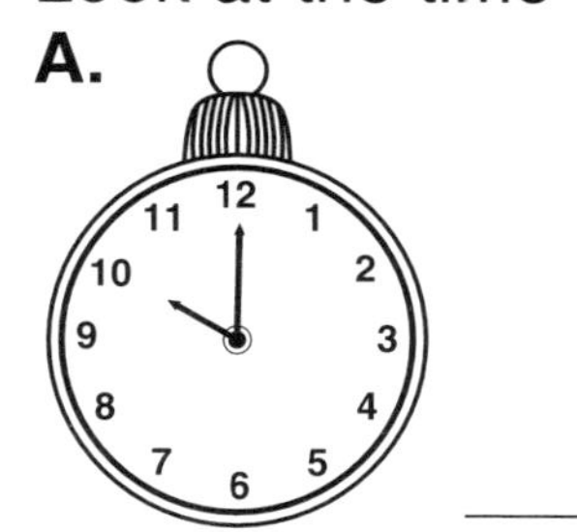
_____ : _____

B. 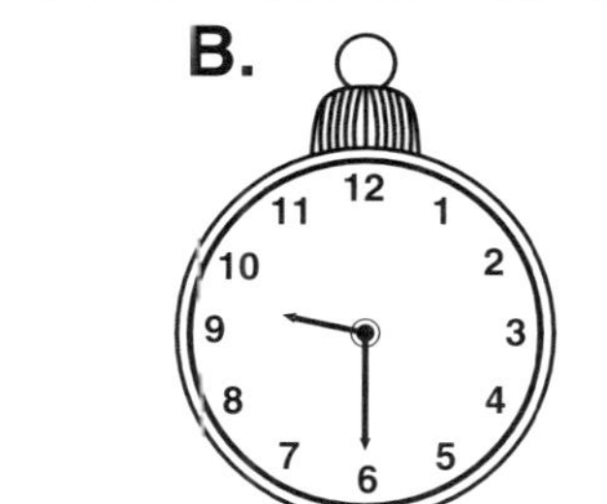
_____ : _____

C. 
_____ : _____

D. 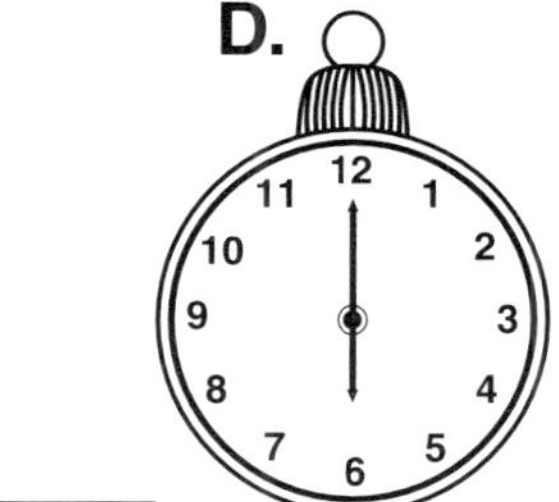
_____ : _____

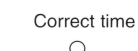

Word Problem

E. Jason's watch is 10 minutes too slow. Jane's watch is 15 minutes too fast. What is the correct time? Draw the hands on the watch to show the correct time.

Jason's watch

Jane's watch

Correct time

Math Problem:

Sentence: ____

Answers: A. 10:00, B. 9:30, C. 1:00 , D. 6:00, E. 3:40

Warm Up

Name: **Date:**

138

Write the time on the line.

A.  ____ : ____

B. 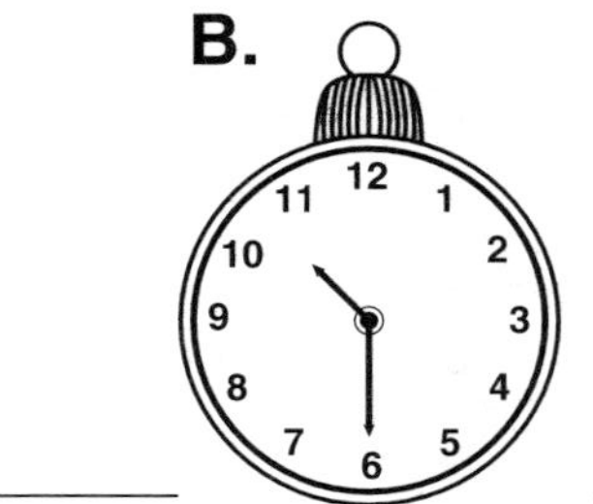 ____ : ____

C.  ____ : ____

D. 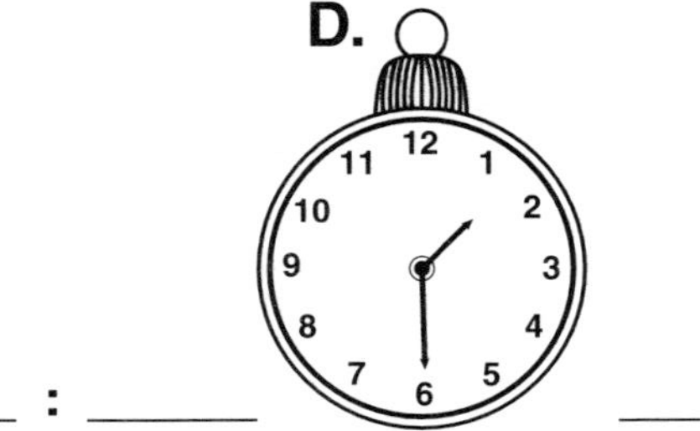 ____ : ____

Word Problem

E. Pauline and her family went to the zoo on Sunday. They spent 10 minutes at the alligator swamp. Then they spent 15 minutes at the camel's sandy habitat. At their last stop, Pauline and her family spent 20 minutes watching the penguins and seals swimming around in the arctic pool. How many minutes did Pauline and her family spend at the zoo?

Math Problem:

Sentence: ____________________

Answers: A. 3:30, B. 10:30, C. 8:30, D. 1:30, E. 45 minutes

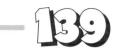

Warm Up

Name: ______________________ Date: ______________________

Look at the time on the clocks. Write the time that is 1/2 hour later.

A.  ____ : ____

B. ____ : ____

C. 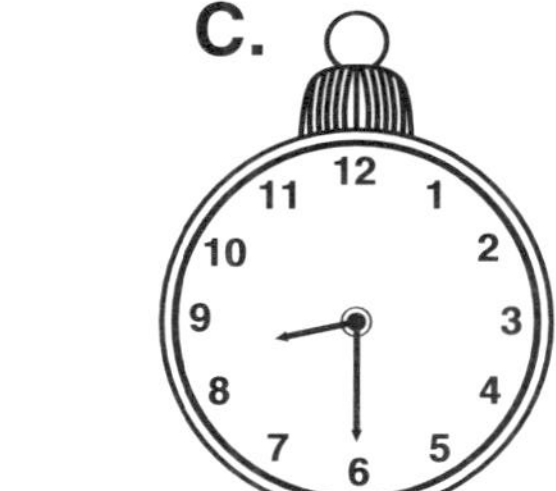 ____ : ____

D. 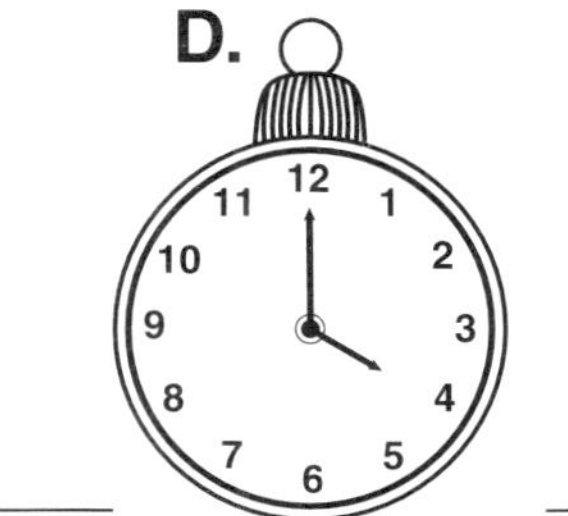 ____ : ____

Word Problem

E. The train leaves the station at 12:05 P.M. It reaches the next station 30 minutes later. What time does the train arrive at the next station?

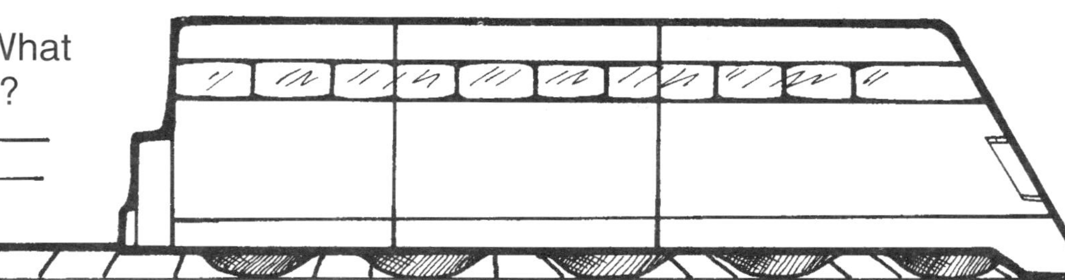

Math Problem:

Sentence: ______________________

Answers: A. 1:30, B. 11:00, C. 9:00, D. 4:30, E. 12:35 P.M.

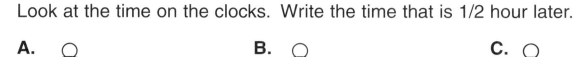

Warm Up

Name: **Date:**

140

Look at the time. Write what time it will be in 10 minutes.

A.

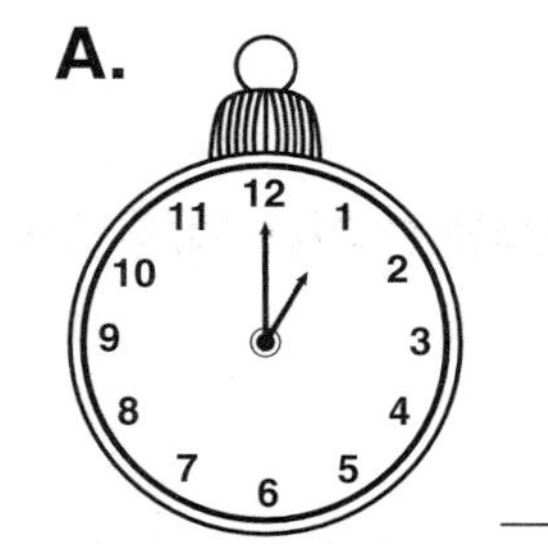

_____ : _____

B.

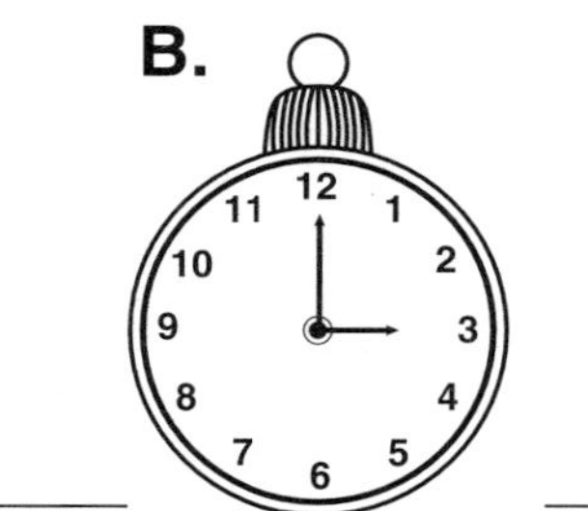

_____ : _____

C.

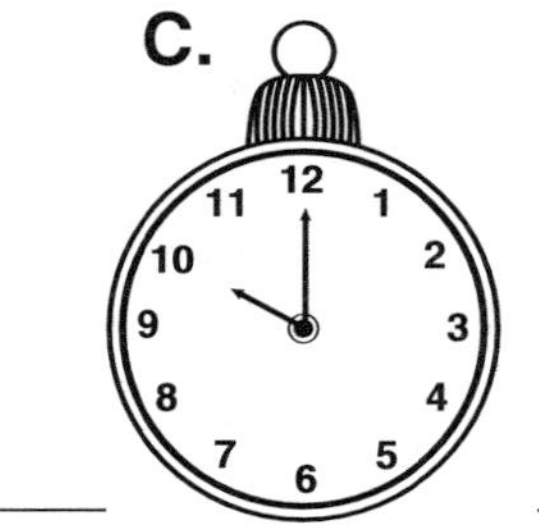

_____ : _____

D.

_____ : _____

Word Problem

E. Nathan was taking a math test. There were 25 math problems on the test. It took Nathan 1 minute to do each math problem. Nathan started the test at 1:00 P.M. What time did Nathan finish the test?

Math Problem:

Sentence: ______________________________

Answers: A. 1:10, B. 3:10, C. 10:10, D. 5:10, E. 1:25 P.M.

Warm Up

Name: Date:

Look at each pair of numbers. Circle the number that is larger.

A. 13 19 **B.** 15 21 **C.** 20 12 **D.** 34 46

Word Problem

E. Mr. Kim grows cherries in his backyard. It takes 30 cherries to make 1 pie. If Mr. Kim wanted to make 3 pies, how many cherries would he need?

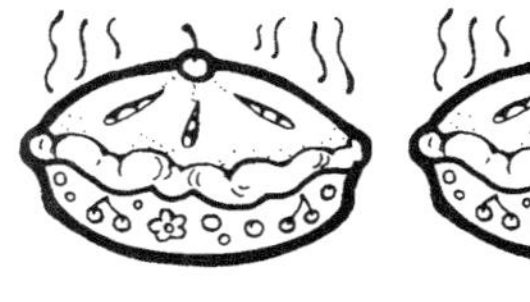

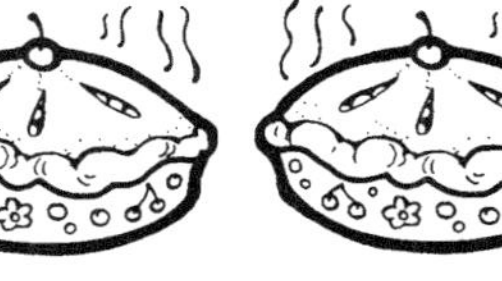

1 pie

_____ cherries

2 pies

_____ cherries

3 pies

_____ cherries

Math Problem:

Sentence: ______________________________

Answers: A. 19, B. 21, C. 20, D. 46, E. 90 cherries

Warm Up

Name: Date:

Make tally marks to show each number. The first one has been done for you.

19	16	12	11
卌 卌 卌 ||||			

Word Problem

Ramona likes to eat sunflower seeds. If she eats 8 sunflowers seeds every day of the week, how many sunflower seeds does she eat in one week? Draw the number of seeds that Ramona ate each day.

Monday:

Tuesday:

Wednesday:

Thursday:

Friday:

Saturday:

Sunday:

Math Problem:

Sentence: ______________________________

Answer: Ramona eats 56 sunflower seeds in one week.

143

Name: ____________ **Date:** ____________

Warm Up

Solve each riddle.

A. Jerome is 14. He is 6 years older than Justine. How old is Justine? ___________

B. Josie planted 15 carrot seeds. She planted 5 more carrot seeds than Carly. How many carrot seeds did Carly plant? ___________

Word Problem

C. There are 3 children in the Hall family. The youngest is Carrie. She is 7 years old. The middle child is Jasper. Jasper is 4 years older than Carrie. The oldest child is Sabrina. Sabrina is 5 years older than Carrie. How old are Jasper and Sabrina?

Math Problem:

Sentence: ____

Answers: A. 8 years old, B. 10 seeds, C. Jasper is 11 years old; Sabrina is 12 years old.

Warm Up

Name: **Date:**

Solve each riddle.

A. Each star has 5 points. How many points are on 5 stars? ____________

B. Each square has 4 corners. How many corners are on 5 squares? ____________

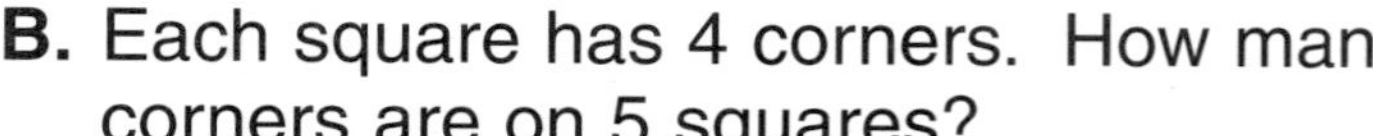 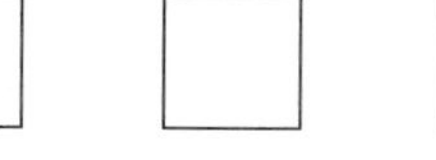 

Word Problem

C. There are 4 castles in the tiny town of Fredstone. Each castle has 16 rooms inside. What are the total number of rooms in Fredstone?

1 castle

_______ rooms

 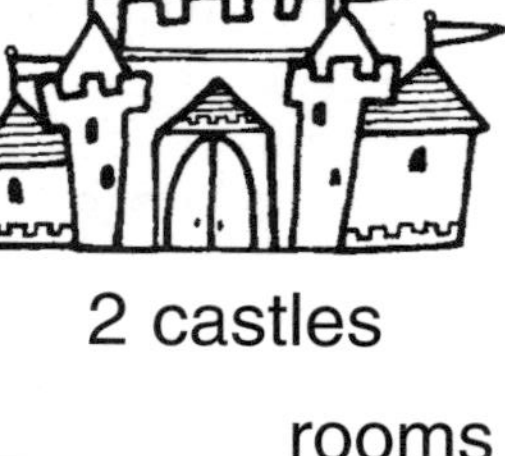

2 castles

_______ rooms

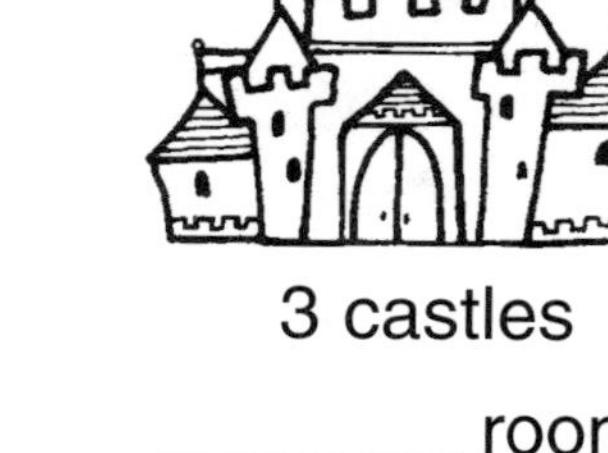

3 castles

_______ rooms

4 castles

_______ rooms

Math Problem:

Sentence: ____

Answers: A. 25 points, B. 20 corners, C. 64 rooms

145

Warm Up

Name: Date:

Count the legs on the cows and sheep. Circle the answer.

A. 0 4

B. 6 12

C. 16 20

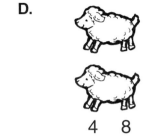

D. 4 8

Word Problem

E. Cowboy Cal takes care of the cows and sheep at the Phoenix Corral. On Monday, he fed 12 sheep. On Tuesday, he fed 14 cows. Each one of the cows and sheep has 4 legs. How many sheep legs are there in all? How many cow legs are there in all?

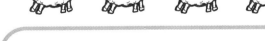

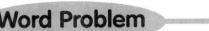

Math Problem:

Sentence: ______________________________

Answers: A. 4, B. 12, C. 16, D. 8, E. There are 48 sheep legs and 56 cow legs.

146

Warm Up

Name: ____________________ Date: ____________________

Add or subtract.

A. 24 + 17 = ______

B. 17 + 4 = ______

C. 7 + 18 = ______

D. 18 + 17 = ______

E. 15 – 6 = ______

F. 25 – 9 = ______

G. 35 – 5 = ______

H. 35 – 10 = ______

Word Problem

I. Count how many triangles you see in the kite below. Remember to count the small and large triangles.

Math Problem:

Sentence: ______________________________

Answers: A. 41, B. 21, C. 25, D. 35, E. 9, F. 16, G. 30, H. 25, I. 11

147

Warm Up

Name: **Date:**

Fact Families: Use the numbers 7, 8, and 15 to make 2 addition problems and 2 subtraction problems.

A. 7 + ________ = ________

C. 15 – ________ = ________

B. 8 + ________ = ________

D. ________ – 8 = ________

Word Problem

E. Each seal can eat 15 fish a day. How many fish can 4 seals eat in one day?

1 seal

____

2 seals

____

3 seals

____

4 seals

____

Math Problem:

Sentence: ________________________________

Answers: A. 7 + 8 = 15, B. 8 + 7 = 15, C. 15 – 7 = 8, D. 15 – 8 = 7, E. 60 fish

Warm Up

Name: Date: **148**

Add.

A. 30 + 13 = ________ **C.** 17 + 23 = ________ **E.** 23 + 14 = ________ **G.** 11 + 17 = ________

B. 17 + 18 = ________ **D.** 10 + 18 = ________ **F.** 22 + 13 = ________ **H.** 15 + 25 = ________

Word Problem

I. Sarah and Jack collect stuffed animals. If Jack has 10 stuffed animals and Sarah has twice as many as Jack, how many stuffed animals does Sarah have?

Math Problem:

Sentence: ________________________________

Answers: A. 43, B. 35, C. 40, D. 28, E. 37, F. 35, G. 28, H. 40, I. 20 stuffed animals

Warm Up

Name: Date:

Solve each math riddle.

A. Sue is 16 years old. Her brother is 2 years older than she is. How old is Sue's brother ? ____________

B. Jake is 27 years old. His sister is 10 years older than he is. How old is Jake's sister ? ____________

Word Problem

C. How many days are there between New Year's Day (January 1) and Martin Luther King Jr.'s birthday (January 15)?

January						
S	M	T	W	Th	F	S
						1
2	3	4	5	6	7	8
9	10	11	12	13	14	15
16	17	18	19	20	21	22
23/30	24/31	25	26	27	28	29

Math Problem:

Sentence: ____

Answers: A. 18 years old, B. 37 years old, C. 13 days

Warm Up

Name: ________ Date: ________ **150**

Subtract.

A. 20 – 13 = ________ **C.** 35 – 4 = ________ **E.** 13 – 10 = ________ **G.** 28 – 7 = ________

B. 17 – 15 = ________ **D.** 20 – 10 = ________ **F.** 32 – 21 = ________ **H.** 20 – 9 = ________

Word Problem

The one-story house has 3 windows, 1 door, and 1 chimney. The two-story house has 5 windows, 1 door, and 2 chimneys. Draw each item on each house and color and decorate both houses. Write your own problem about the differences between the two houses.

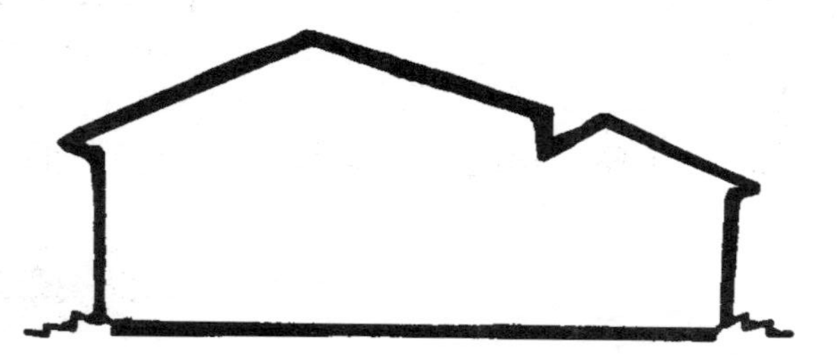

Math Problem:

Sentence: ________________________

Answers: A. 7, B. 2, C. 31, D. 10, E. 3, F. 11, G. 21, H. 11

Name: Date:

151

Warm Up

Add.

A. 18 + 25 = ______ C. 24 + 15 = ______ E. 16 + 13 = ______ G. 14 + 18 = ______

B. 22 + 15 = ______ D. 16 + 35 = ______ F. 33 + 13 = ______ H. 27 + 12 = ______

Word Problem

I. One starfish has 5 arms. How many arms would 4 starfish have in all?

1 starfish

______ arms

2 starfish

______ arms

3 starfish

______ arms

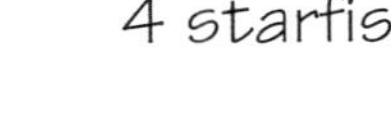

4 starfish

______ arms

Math Problem:

Sentence: ______________________________

Answers: A. 43, B. 37, C. 39, D. 51, E. 29, F. 46, G. 32, H. 39, I. 20 arms

Warm Up

Name: ____________________ Date: ____________________

152

Add.

A. 19 + 18 = ________ **C.** 12 + 17 = ________ **E.** 10 + 31 = ________ **G.** 38 + 12 = ________

B. 16 + 30 = ________ **D.** 20 + 19 = ________ **F.** 30 + 12 = ________ **H.** 34 + 14 = ________

Word Problem

I. A dragonfly has 4 wings. If there are 14 dragonflies flying in the sky, how many wings are there in all?

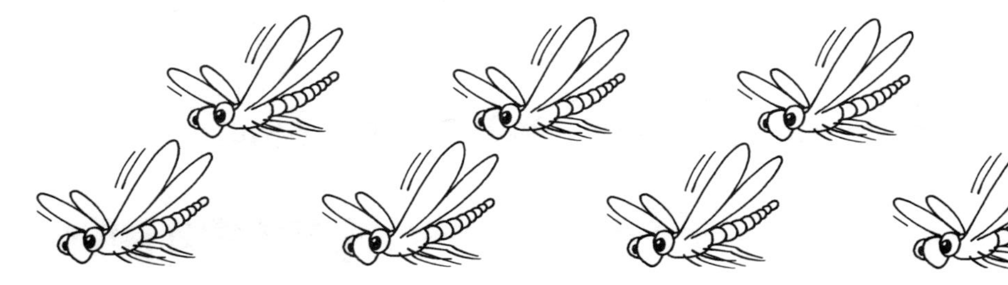

Math Problem:

Sentence: ____________________

Answers: A. 37, B. 46, C. 29, D. 39, E. 41, F. 42, G. 50, H. 48, I. 56 wings

Name: **Date:**

153

Warm Up

Subtract.

A. 27 – 5 = ________ **C.** 34 – 14 = ________ **E.** 28 – 19 = ________ **G.** 31 – 11 = ________

B. 19 – 16 = ________ **D.** 35 – 8 = ________ **F.** 22 – 9 = ________ **H.** 16 – 7 = ________

Word Problem

I. Maria had 8 pears. Her sister gave her 12 more pears. Her brother gave her some pears, too. Now Maria has two dozen pears. Color the pears that Maria had green. Color the pears that her sister gave her yellow. Color the pears that her brother gave her red. How many pears did her brother give to her? (remember that one dozen = 12)

Math Problem:

Sentence: ____________________

Answers: A. 22, B. 3, C. 20, D. 27, E. 9, F. 13, G. 20, H. 9, I. 4 pears

Warm Up

Name: **Date:** **154**

Subtract.

A. 35 – 16 = ________ **C.** 53 – 8 = ________ **E.** 49 – 13 = ________ **G.** 57 – 18 = ________

B. 46 – 15 = ________ **D.** 44 – 23 = ________ **F.** 38 – 20 = ________ **H.** 23 – 13 = ________

Word Problem

I. There are 45 students on 3 school buses. There are 15 students on the first bus, and 20 students are on the second bus. The remaining students are on the third bus. How many students are on the third bus?

bus #1

15 students

bus #2

20 students

bus #3

____ students

Math Problem:

Sentence: ______________________________

Answers: A. 19, B. 31, C. 45, D. 21, E. 36, F. 18, G. 39, H. 10, I. 10 students

Warm Up

Name: Date: **155**

Add.

A. 20 + 40 = ________ **C.** 17 + 47 = ________ **E.** 50 + 20 = ________ **G.** 36 + 46 = ________

B. 12 + 36 = ________ **D.** 23 + 15 = ________ **F.** 14 + 24 = ________ **H.** 41 + 14 = ________

Word Problem

I. The 3 polar bears at the local zoo each eat a dozen fish a day. How many fish would these bears eat in 1 day? (remember that one dozen = 12)

1 polar bear ________ 2 polar bears ________

3 polar bears ________

Math Problem:

Sentence: ________________________________

Answers: A. 60, B. 48, C. 64, D. 38, E. 70, F. 38, G. 82, H. 55, I. 36 fish

Warm Up

Name: **Date:**

156

Count by 2's to fill in the missing numbers.

A. 0, 2, ___, ___, ___, 10, ___, 14, ___, ___, 20, ___, 24, ___, 28, 30, ___, ___, 36, ___, ___, 42, 44, ___

Word Problem

B. Jimmy has 2 dozen marbles. There are 4 red marbles, 10 green marbles, and the rest are blue. How many marbles are blue? (remember that one dozen = 12)

Math Problem:

Sentence: ____

Answers: A. 4, 6, 8, 12, 16, 18, 22, 26, 32, 34, 38, 40, 46, B. 10 marbles

Warm Up **Name:** **Date:** **157**

Count by 4's to fill in the missing numbers.

A. 0, 4, ___, ___, 16, ___, ___, ___, 32, ___, 40, ___, ___, 52, 56, ___

Word Problem

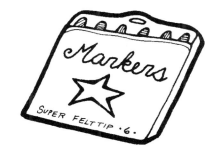

B. Megan bought a box of crayons for $3.25. She also bought a set of markers for $4.95. If she took $10 with her, how much change did Megan get back?

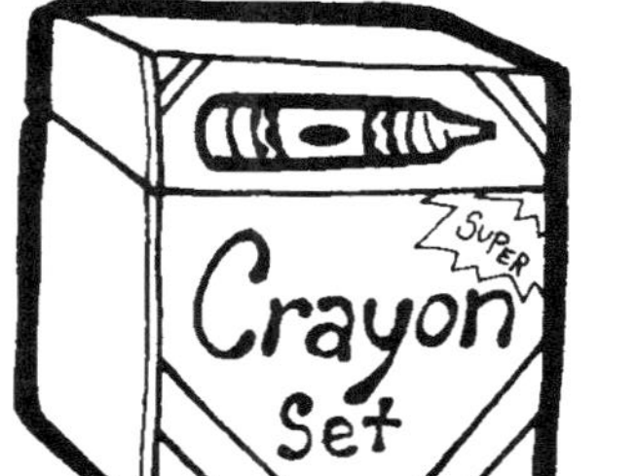

Math Problem:

Sentence: ______________________________

Answers: A. 8, 12, 20, 24, 28, 36, 44, 48, 60, B. $1.80

158

Name: **Date:**

Warm Up

Draw a line to cut each shape into two equal parts.

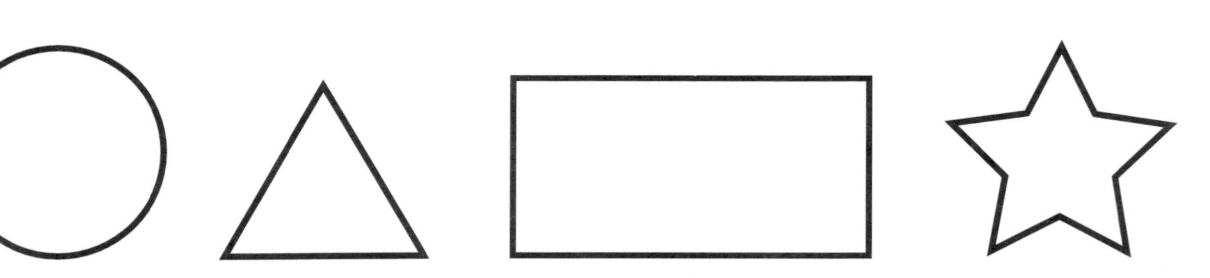

Word Problem

The snail and the rabbit had a race. If it took the rabbit 15 minutes to finish the race and it took the snail 25 minutes more, how long did it take the snail to finish the race?

Math Problem:

Sentence: ______________________________

Answer: 40 minutes

Warm Up

Name: Date:

159

Count by 5's. Write the numbers on the lines.

A.

_____ _____ _____ _____ _____ _____ _____ _____ _____ _____

Word Problem

B. When Micah returned home from school, he took out all of his nickels from his pocket. How much money does Micah have?

Math Problem:

Sentence: ______________________________

Answers: A. 5, 10, 15, 20, 25, 30, 35, 40, 45, 50, B. 45¢

Name: ____________ Date: ____________

160

Warm Up

Count by 5's to fill in the missing numbers.

A. 0, 5, ________, 15, ________, 25, ________, 35, ________, 45, ________, 55, ________

Word Problem

B. Each turkey has 5 long tail feathers. Draw the tail feathers for each turkey and count the total number of feathers on the 6 turkeys.

Math Problem:

Sentence: ____

Answers: A. 10, 20, 30, 40, 50, 60, B. 30 tail feathers

161

Warm Up

Name: ____________________ **Date:** ____________________

Count by 5's to fill in the missing numbers.

A. 25, ________, ________, 40, ________, ________, 55, ________, ________, 70, ________

Word Problem

B. Betty saw some bats flying out of a cave last night. Each cave can hold 15 bats. How many bats can 2 caves hold?

Math Problem:

Sentence: ____

Answers: A. 30, 35, 45, 50, 60, 65, 75, B. 30 bats

162

Warm Up

Name: **Date:**

Count by 5's to fill in the missing numbers to 100.

A. 50, ________, ________, ________, 70, ________, ________, 85, ________, ________, 100

Word Problem

B. On cold winter 4 days, Kevin likes to make vegetable soup. Kevin puts 5 big tomatoes, 3 green peppers, 4 potatoes, and carrots into the pot. How many vegetables does Kevin use to make the soup? Using the counters provided on page 256, glue the correct number of each vegetable used to make the soup.

Math Problem:

Sentence: ____

Answers: A. 55, 60, 65, 75, 80, 90, 95, B. 16 vegetables

Warm Up **Name:** ____________ **Date:** ____________ **163**

Count by 10's to fill in the missing numbers

A. 0, ________, ________, 30, ________, ________, 60, ________, 80, ________, 100

Word Problem

B. Violet emptied out her piggy bank to count the change. Help Violet sort her change into groups of 50¢. Write a math problem and a sentence about the different ways 50¢ can be made.

Math Problem:

Sentence: ____

Answers: A. 10, 20, 40, 50, 70, 90, B. Answers will vary.

Warm Up

Name: Date:

164

Count by 10's to fill in the missing numbers

A. 0, 10, ________, 30, ________, 50, ________, 70, ________, 90, ________

Word Problem

B. One ant has 6 legs. How many arms would 6 ants have?

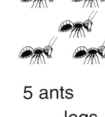

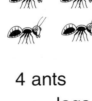

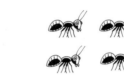

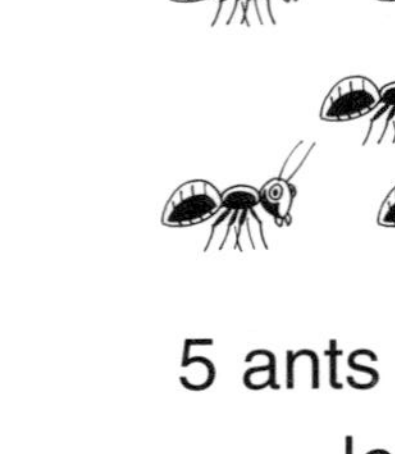

1 ant	2 ants	3 ants	4 ants	5 ants	6 ants
_____ legs	_____ legs	_____ legs	_____ legs	_____ legs	_____ legs

Math Problem:

Sentence: ____

Answers: A. 20, 40, 60, 80, 100, B. 36 legs

Warm Up

Name: Date: **165**

Write the number of ones. The first one has already been done for you.

A. 4

B. ______

C. ______

D. ______

Word Problem

E. The Wilson family receives a lot of mail! On Monday, the Wilsons received 7 letters. On Friday, the Wilsons received 5 letters. On Saturday, the Wilsons received 3 letters. How many letters did the Wilsons receive in a week?

Monday: ______ Friday: ______ Saturday: ______

Math Problem:

Sentence: ______

Answers: A. 4, B. 5, C. 1, D. 2, E. 15 letters

Warm Up

Name: Date:

166

Write the number of ones. The first one has already been done for you.

A. 6

B. ______

C. ______

D. ______

Word Problem

E. At the park there is a pond full of ducks. There are 10 ducks in the pond. Count the total number of wings and feet of all the ducks in the pond. (Remember, each duck has 2 wings and 2 feet.)

Math Problem:

Sentence: ______

Answers: A. 6, B. 7, C. 5, D. 8, E. 20 wings and 20 feet

Warm Up

Name: Date: **167**

Write the number of tens. The first one has already been done for you.

A. 2 tens

B. _______ tens

C. _______ ten

D. _______ tens

Word Problem

E. Mindy and Suzy spent the afternoon making cupcakes. Mindy made 13 cupcakes. If 25 cupcakes were made how many did Suzy bake? Complete the chart to show how many cupcakes each person baked.

Mindy																		
Suzy																		

Math Problem:

Sentence: ______________________________

Answers: A. 2 tens, B. 3 tens, C. 1 ten, D. 4 tens E. 12 cupcakes

Name: Date:

168

Warm Up

Write the number of tens. The first one has already been done for you.

A. 5 tens

B. ______ tens

C. ______ tens

D. ______ tens

Word Problem

E. How many different ways can you put the 10 cookies into 2 jars?

jar #1		jar #2	
5 cookies	+	5 cookies	= 10 cookies
________	+	________	= 10 cookies
________	+	________	= 10 cookies
________	+	________	= 10 cookies
________	+	________	= 10 cookies

Math Problem:

Sentence: ________________________________

Answers: A. 5 tens, B. 6 tens, C. 7 tens, D. 3 tens, E. Answers will vary.

169

Warm Up

Name: ____________ Date: ____________

Write the number of tens and ones. The first one has already been done for you.

A. 2 tens 3 ones

B. ____ tens ____ ones

C. ____ tens ____ ones

D. ____ tens ____ ones

Word Problem

E. Bryan and Trevor were collecting different types of leaves for their science project. They found 4 leaves on Monday. On Tuesday they found twice as many leaves as they did on Monday. On Wednesday they found 10 leaves. How many leaves did Bryan and Trevor find each day, and what was the total number of leaves that they found for their science project?

Monday: ____________

Tuesday: ____________

Wednesday: ____________

Total number of leaves: ____________

Math Problem:

Sentence: ____________

Answers: A. 2 tens, 3 ones, B. 3 tens, 2 ones, C. 3 tens, 0 ones, D. 4 tens, 0 ones, E. Monday-4 leaves, Tuesday-8 leaves, Wednesday-10 leaves; Total number of leaves = 22 leaves

Warm Up

Name: ______ **Date:** ______

170

Write the number of tens and ones. The first one has already been done for you.

A. 6 tens 2 ones

B. ____ tens ____ ones

C. ____ tens ____ ones

D. ____ tens ____ ones

Word Problem

E. Connie took a poll for a class assignment. She asked 20 classmates, "Do you like pumpkin or apple pie?" If 8 of her classmates selected apple pie, how many of Connie's classmates selected pumpkin pie? Draw the exact number of pies to represent the results from Connie's pie poll in the graph below.

(apple)	
(pumpkin)	

Math Problem:

Sentence: ____________________

Answers: A. 6 tens, 2 ones, B. 5 tens, 6 ones, C. 4 tens, 3 ones, D. 3 tens, 7 ones, E. 12 classmates selected pumpkin pie

Warm Up

Name: Date:

Look at each number. Draw the number of tens and ones. The first one has already been done for you.

A.	B.	C.	D.
15	22	40	65

Word Problem

Using a kitchen timer or stopwatch, give yourself one minute to draw as many stars as possible. After the minute is up, circle the stars into groups of tens. How many groups of ten did you make? How many ones are left? Write a math problem and a sentence about the number of stars you made.

Math Problem:

Sentence: ______________________________

Name: Date:

172

Warm Up

Write the number. The first one has already been done for you.

A. 3 tens 7 ones

37

B. 8 tens 0 ones

____

C. 1 ten 9 ones

____

D. 5 tens 6 ones

____

Word Problem

E. Eric has a lot of posters about different dinosaurs. He bought 4 new posters. Each one costs $5. How much did Eric spend on the 4 dinosaur posters?

Math Problem:

Sentence: ____________________

Answers: A. 37, B. 80, C. 19, D. 56, E. $20

173

Warm Up

Name: **Date:**

Use the numbers in each set to make the largest 3-digit number possible.

A. 4, 0, 1 __________

B. 2, 3, 6 __________

C. 5, 9, 7 __________

D. 2, 8, 1 __________

Word Problem

E. Measure the different lengths of the objects below. Which object is longer?

Math Problem:

Sentence: ____________________

Answers: A. 410, B. 632, C. 975, D. 821, E. whale

Name: Date:

Warm Up

Use the numbers in each set to make the smallest 3-digit number possible.

A. 1, 4, 6 ________ **B.** 9, 3, 7 ________ **C.** 1, 5, 2 ________ **D.** 3, 5, 1 ________

Word Problem

Dot the Dalmatian is covered in spots.

E. Clue #1: Dot has more than 50 spots.
Clue #2: Dot has fewer than 70 spots.
Clue #3: Dot has an even number of spots.
Clue #4: When you count by 5's and by 10's you say the number.
Can you figure out how many spots Dot has?

Math Problem:

Sentence: ________________________________

Answers: A. 146, B. 379, C. 125, D. 135, E. 60 spots

Warm Up

Name: Date:

175

Look at each number. Answer the questions.

A. Circle the number in the tens place.

1 5 8

B. Circle the number in the ones place.

2 9 7

C. Circle the number in the hundreds place.

3 4 6

D. Write the largest number possible using 2, 8, and 1.

____

Word Problem

E. Using the numbers 2, 5, and 8, what are the different numbers you can make?

numbers with 1 digit	numbers with 2 digits	numbers with 3 digits

I made __________ numbers using the digits 2, 5, and 8.

Math Problem:

Sentence: ____

Answers: A. 5, B. 7, C. 3, D. 821, E. Answers will vary.

Warm Up

Name: Date:

Look at each number. Answer the questions.

A. Circle the number in the ones place.

1 0 6

B. Circle the number in the hundreds place.

9 8 1

C. Circle the number in the tens place.

2 1 1

D. Make the smallest number possible using 2, 8, and 1.

____

Word Problem

E. Look at each group of items. How are the items sorted?

What do all 3 groups of items have in common? __________

Math Problem:

Sentence: __________

Answers: A. 6, B. 9, C. 1, D. 128, E. group #1: fruits, group #2: vegetables, group #3: plants; they all grow.

177

Warm Up

Name: Date:

Subtract.

A. $\begin{array}{r} 71 \\ -\ 11 \\ \hline \end{array}$

B. $\begin{array}{r} 79 \\ -\ 29 \\ \hline \end{array}$

C. $\begin{array}{r} 56 \\ -\ 33 \\ \hline \end{array}$

D. $\begin{array}{r} 34 \\ -\ 20 \\ \hline \end{array}$

Word Problem

E. There are 10 students in Ms. Chang's class. Each student in Ms. Chang's class planted 4 seeds. How many seeds did the class plant?

Math Problem:

Sentence: ____________________

Answers: A. 60, B. 50, C. 23, D. 14, E. 40 seeds

Warm Up

Name: ______________________ Date: ______________________

178

Count by 5's to fill in the missing numbers.

A. 5, ______, ______, ______, ______, 30, ______, ______, ______, ______, 55, ______,

______, ______, ______, ______, ______, ______, ______, 100

Word Problem

B. Alicia has lots of flowers in her garden. She has the following types of flowers in her garden: tulip, rose, and daisy. She wants to plant 5 of each flower. How many flowers will Alicia plant in her garden? Using the counters provided on page 256, glue the correct number of each flower.

Math Problem:

Sentence: ____

Answers: A. 10, 15, 20, 25, 35, 40, 45, 50, 60, 65, 70, 75, 80, 85, 90, 95, B. 15 flowers

179

Warm Up

Name: Date:

Count by 10's to 100.

A. 0, ________, ________, ________, ________, ________, ________, ________, ________, ________, ________

Word Problem

B. Cory asked some classmates if they liked green grapes or purple grapes. He made a graph of their answers. What information can you discover by looking at the graph?

green grapes

purple grapes

Math Problem:

Sentence: ________________________________

Answers: A. 10, 20, 30, 40, 50, 60, 70, 80, 90, 100, B. 8 people like green grapes and 6 people like purple grapes.

Warm Up

Name: Date: **180**

Add.

A. $30 + 15 = $

B. $50 + 29 = $

C. $1 + 10 = $

D. $44 + 44 = $

Word Problem

E. Using circles, draw a set of eyes for each gingerbread man. Draw 2 triangled-shaped buttons for each gingerbread man. How many circles did you draw? How many triangles did you draw?

Math Problem:

Sentence: ______________________________

Answers: A. 45, B. 79, C. 11, D. 88, E. 10 circles and 10 triangles

Warm Up

Name: ______________________ Date: ______________________

Add.

A.	B.	C.	D.
34 + 20	25 + 31	61 + 26	18 + 70

Word Problem

E. Grandpa made a dozen of his special chocolate chip cookies. Grandma ate 3 of the cookies. Kendra ate 4 of the cookies. Kelvin ate 2 of the cookies. How many of Grandpa's special chocolate chip cookies are left? Cross out all the cookies that have already been eaten.

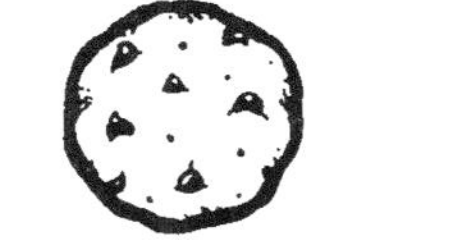
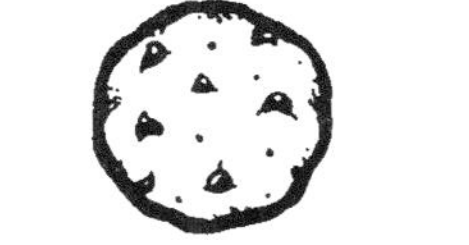
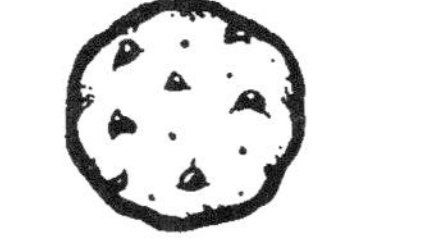

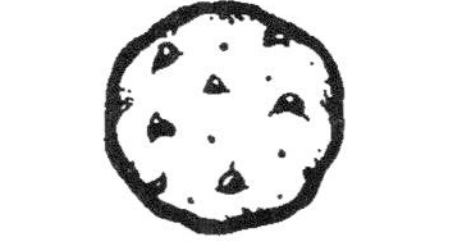
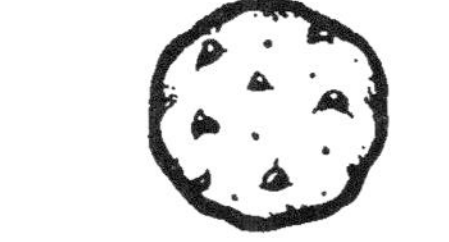

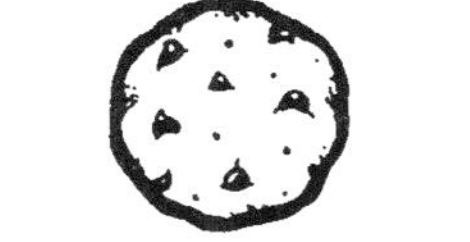

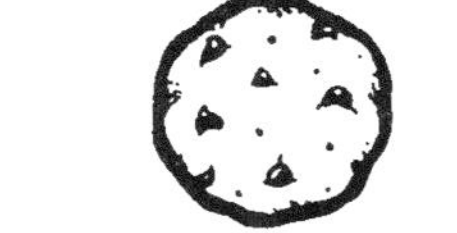

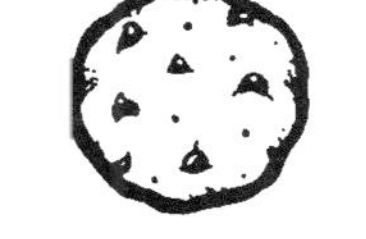

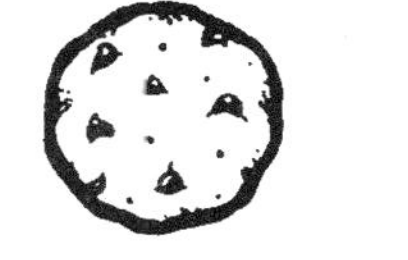

Math Problem:

Sentence: ______________________

Answers: A. 54, B. 56, C. 87, D. 88, E. 3 cookies

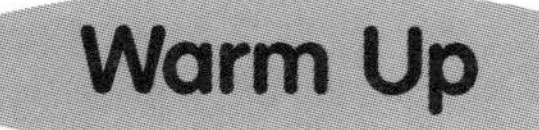

Warm Up

Name: **Date:**

Subtract.

A. $55 - 15$

B. $81 - 30$

C. $16 - 6$

D. $40 - 20$

Word Problem

E. Faith and Brad planted 4 sunflower plants. Each sunflower had 8 sunflower seeds on it. How many sunflower seeds were there in all?

1 sunflower

_____ seeds

2 sunflowers

_____ seeds

3 sunflowers

_____ seeds

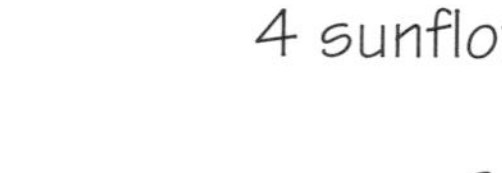

4 sunflowers

_____ seeds

Math Problem:

Sentence: ____________________

Answers: A. 40, B. 51, C. 10, D. 20, E. 32 sunflower seeds

Warm Up

Name: Date:

Add or subtract.

A. 40 − 5

B. 50 + 15

C. 99 − 37

D. 61 − 30

Word Problem

E. The Parker family went to the grocery store. They bought 1 bag of cherries, 10 apricots, and 4 ears of corn. How much did they spend at the store?

bag of cherries $2.15

5 apricots for $1.85

4 ears of corn for $1.25

Math Problem:

Sentence: ____________________

Answers: A. 35, B. 65, C. 62, D. 31, E. $7.10

Warm Up

Name: Date:

184

Add or subtract.

A. 74 + 13 = ____

B. 92 + 2 = ____

C. 31 − 11 = ____

D. 83 − 60 = ____

Word Problem

E. Captain Johnson has a big fishing boat. There are 87 seats on the boat. There are 24 empty seats. How many seats are filled?

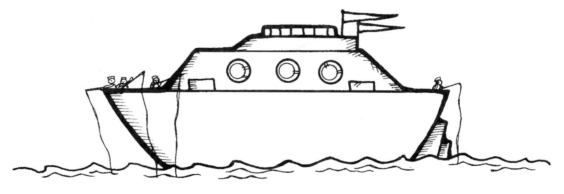

Math Problem:

Sentence: ____________________

Answers: A. 87, B. 94, C. 20, D. 23, E. 63 seats

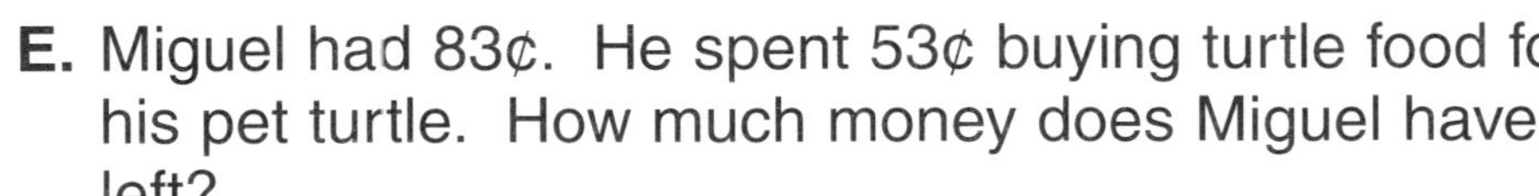

Warm Up

Name: **Date:**

Subtract.

A. 20¢ − 13¢

B. 67¢ − 57¢

C. 39¢ − 11¢

D. 98¢ − 35¢

Word Problem

E. Miguel had 83¢. He spent 53¢ buying turtle food for his pet turtle. How much money does Miguel have left?

Math Problem:

Sentence: ____________________

Answers: A. 7¢, B. 10¢, C. 28¢, D. 63¢, E. 30¢

Warm Up

Name: **Date:**

Count the pennies. Write the total on the line.

A.  ____________

B.  ____________

C.  ____________

D.  ____________

Word Problem

E. The lollipop costs 25¢. Does Henrietta have enough money to buy the lollipop? How much more money does she need? Using the counters provided on page 256, glue the correct number of coins.

Math Problem:

Sentence: ____

Answers: A. 3¢, B. 5¢, C. 6¢, D. 4¢, E. No, 10¢ more (10 pennies)

Warm Up

Name: **Date:**

187

Count the money. Write the total on the line.

A.

____

B.

____

C.

____

D.

____

Word Problem

E. One afternoon, Leticia decided to count the money in her piggy bank. How much money did Leticia have in her piggy bank?

Math Problem:

Sentence: ____

Answers: A. 8¢, B. 10¢, C. 5¢, D. 10¢, E. 52¢

Warm Up

Name: ______________________ **Date:** ______________________ **188**

Count the money. Write the total on the line.

A. ________

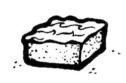

B. ________

C. ________

D. ________

Word Problem

E. Hilda's mother gives her $2 for lunch money every day. List what Hilda can buy for lunch today.

2 cookies 50¢ | milk 50¢ | orange juice box 35¢ | apple 30¢ | banana 40¢ | brownie 55¢

Math Problem:

Sentence: ____

Answers: A. 4¢, B. 5¢, C. 5¢, D. 10¢, E. Answers will vary.

Warm Up

Name: **Date:**

Count the money. Write the total amount on the line.

A.

______

B.

______

C.

______

D.

______

Word Problem

How much is your first name worth? Use the chart below to figure out the value of your first name.

Vowels 5¢ each
a, e, i, o, u
Consonants 10¢ each
b, c, d, f, g, h, j, k, l, m, n, p, q, r, s, t, v, w, x, y, z

Name: ___ ___ ___ ___ ___ ___ ___ ___ ___ ___

Value: ______________________ = ______¢

Math Problem:

Sentence: ______________________

Answers: A. 5¢, B. 6¢, C. 10¢, D. 7¢

Warm Up

Name: ______________________ Date: ______________________

Count the money. Write the total on the line.

A.

____

B.

____

C.

____

D.

____

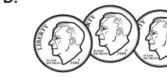

Word Problem

Count the total amount of money shown and calculate how much the toy airplane costs.

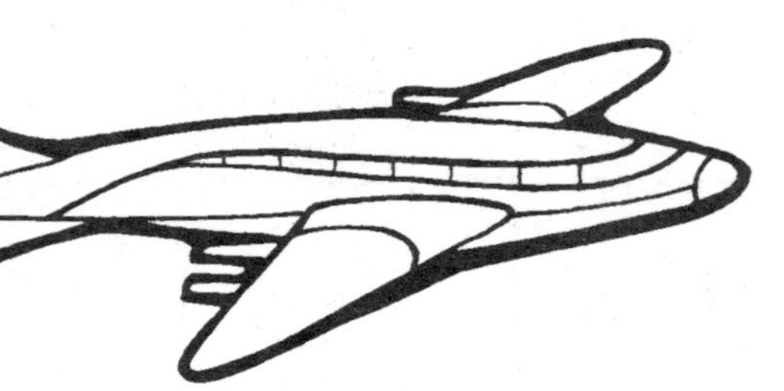

Math Problem:

Sentence: ____

Answers: A. 10¢, B. 40¢, C. 20¢, D. 30¢, E. $2.30

Counters/Manipulatives

Use with page 64.

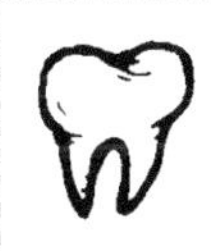 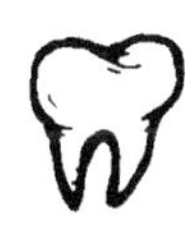 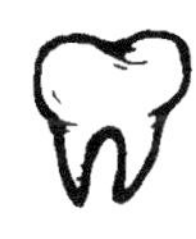 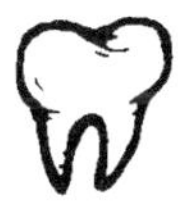 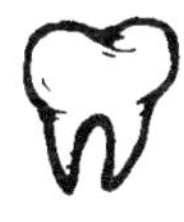 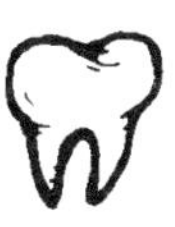 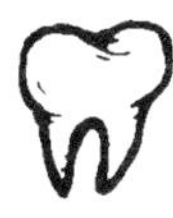 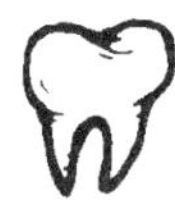 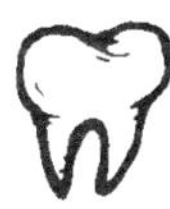 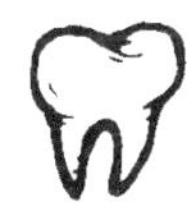

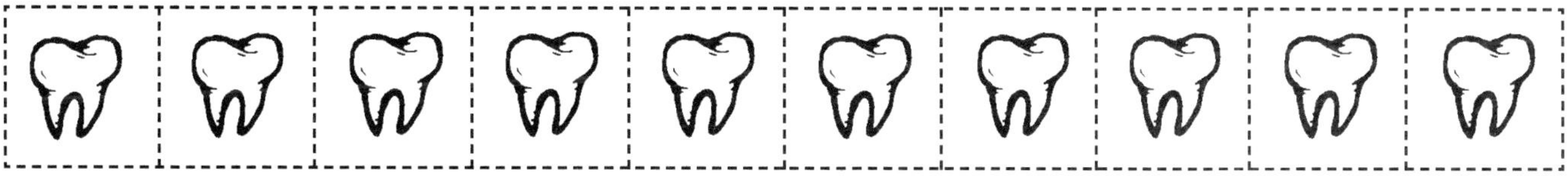

Use with page 72.

Use with page 75.

Use with page 86.

    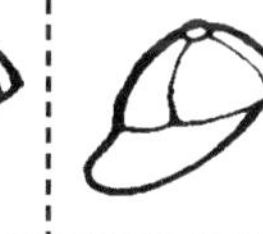 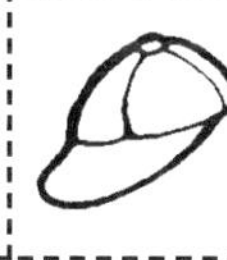    

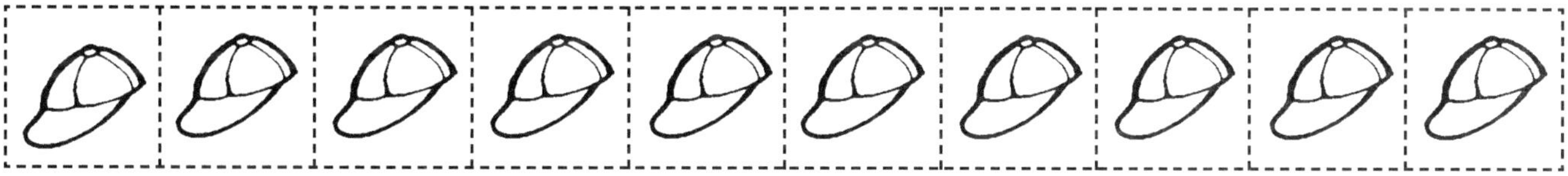

Use with page 87.

Use with page 93.

Use with page 95.

Use with page 95.

Counters/Manipulatives *(cont.)*

Use with page 97.

Use with page 97.

Use with page 101.

Use with page 102.

Use with page 113.

Use with page 113.

Use with page 114.

Use with page 114.

Counters/Manipulatives *(cont.)*

Use with page 118.

Use with page 118.

Use with page 119.

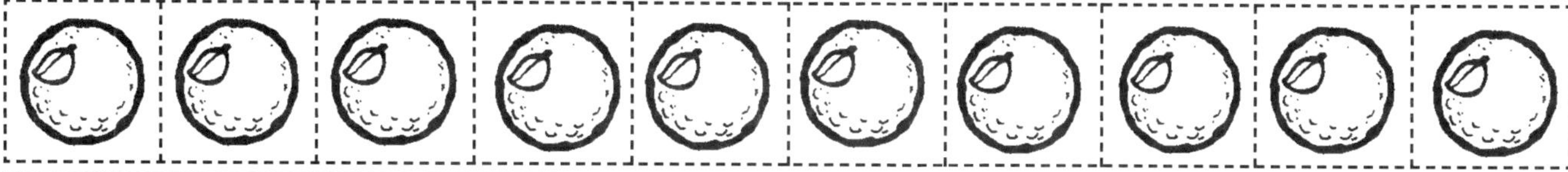

Use with page 119.

Use with page 119.

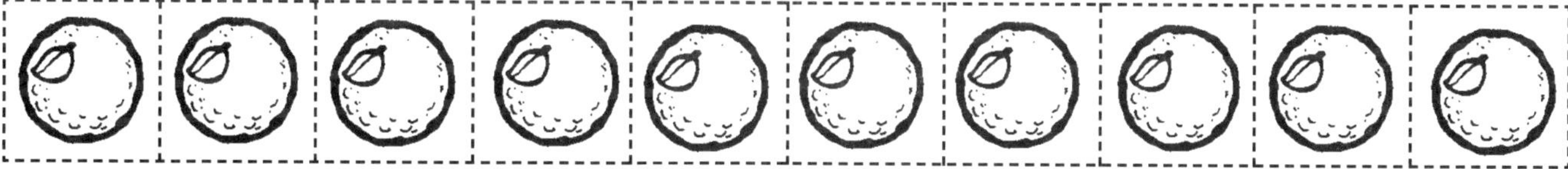

Use with page 123.

Use with page 123.

Use with page 123.

Counters/Manipulatives *(cont.)*

Use with page 129.

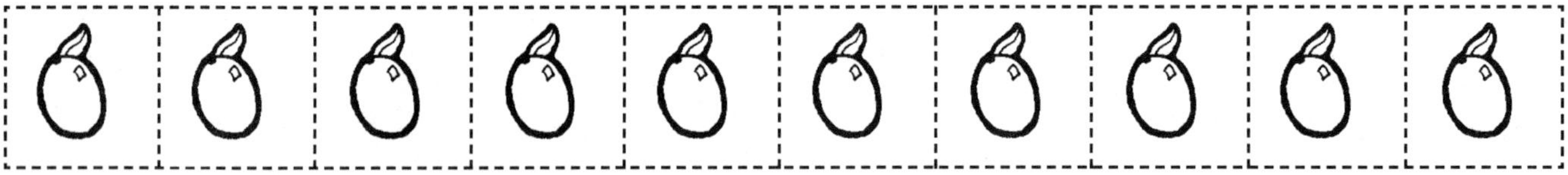

Use with page 132.

Use with page 132.

Use with page 138.

Use with page 138.

Use with page 138.

Use with page 141.

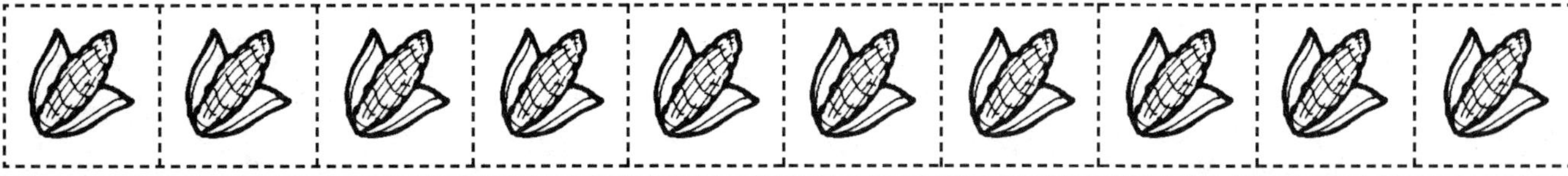

Use with page 141.

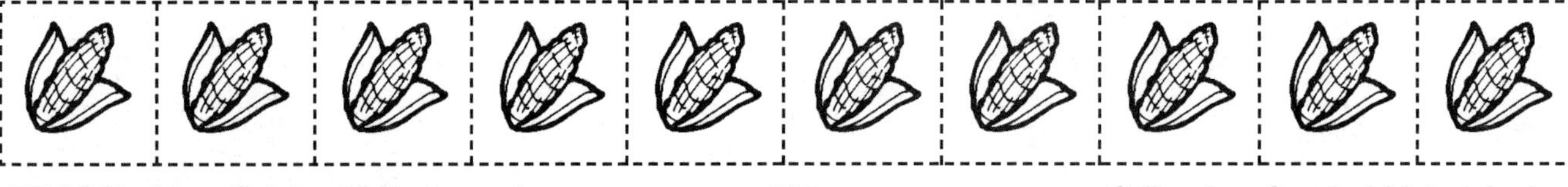

Counters/Manipulatives *(cont.)*

Use with page 146.

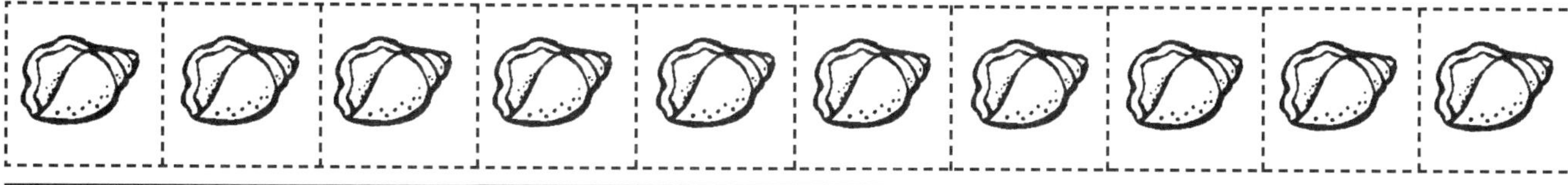

Use with page 160.

Use with page 160.

Use with page 171.

Use with page 171.

Use with page 175.

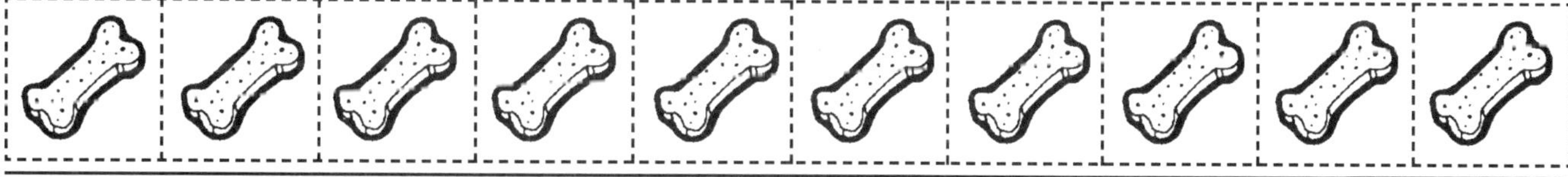

Use with page 176.

Use with page 176.

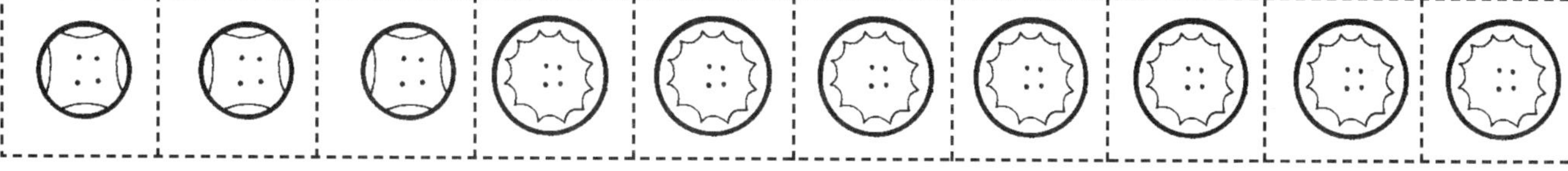

Counters/Manipulatives *(cont.)*

Use with page 188.

Use with page 222.

Use with page 222.

Use with page 223.

Use with page 238.

Use with page 238.

Use with page 238.

Use with page 246.

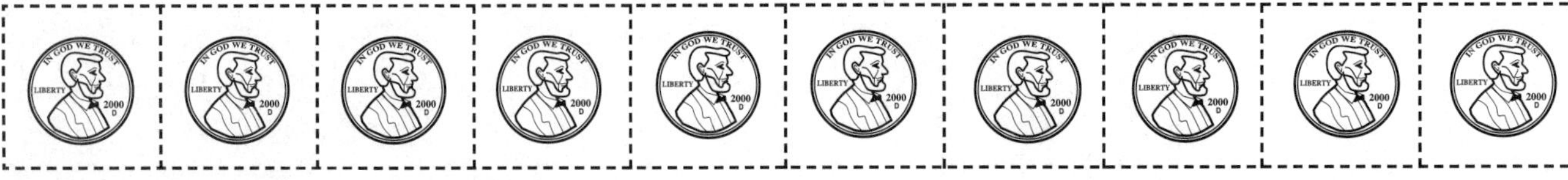

